Adolf Grünbaum

Essays on Actions and Events

Essays on Actions and Events

DONALD DAVIDSON

CLARENDON PRESS · OXFORD

Oxford University Press, Walton Street, Oxford OX2 6DP

Oxford New York Toronto
Delhi Bombay Calcutta Madras Karachi
Petaling Jaya Singapore Hong Kong Tokyo
Nairobi Dar es Salaam Cape Town
Melbourne Auckland
and associated companies in
Berlin Ibadan

Oxford is a trade mark of Oxford University Press

Published in the United States
by Oxford University Press, New York

First published 1980
Reprinted with corrections 1982, 1985, 1986, 1989, 1990, 1991

British Library Cataloguing in Publication Data

Davidson, Donald
Essays on actions and events.
1. Act (Philosophy)
I. Title
128'.3   B105.A35   80–40064
ISBN 0–19–824529–7
ISBN 0–19–824637–4 (Pbk)

Printed in Great Britain by Biddles Ltd
Guildford and King's Lynn

Contents

Provenance of the Essays and Acknowledgements

Essay 1, 'Actions, Reasons, and Causes', was presented in a symposium on 'Action' at the 1963 meeting of the American Philosophical Association and published in the *Journal of Philosophy* 60 (1963). It is reprinted by permission of the editors.

Essay 2, 'How is Weakness of the Will Possible?', was first published in *Moral Concepts*, edited by Joel Feinberg, Oxford Readings in Philosophy, 1970. Drafts of the paper had been read at the 1967 Annual Oregon Colloquium in Philosophy, where I had the benefit of the careful comments of Donald G. Brown, and at the Chapel Hill Colloquium in Philosophy, where I heard Gilbert Harman's skilful criticisms.

Essay 3, 'Agency', was presented at the fourth philosophy colloquium held at the University of Western Ontario in November 1968, and was published in *Agent, Action, and Reason*, edited by Robert Binkley, Richard Bronaugh, and Ausonio Marras, University of Toronto Press, 1971. James Cornman commented on the paper at the colloquium, and the printed version gained by his advice. It is reprinted by permission of the University of Toronto Press.

The fourth Essay, 'Freedom to Act', was first published in *Essays on Freedom of Action*, edited by Ted Honderich, Routledge and Kegan Paul, 1973 (reprinted as a paperback in 1978).

Essay 5, 'Intending', was read at the first Jerusalem Philosophical Encounter, December 1974, where Stuart Hampshire made a spirited response. My paper and his response were printed in *Philosophy of History and Action*, edited by Yirmiaku Yovel, D. Reidel and The Magnes Press, The Hebrew University, 1978. An earlier version was presented at the Chapel Hill Colloquium in Philosophy in October 1974, where it received a thorough going over by Paul Grice.

'The Logical Form of Action Sentences', Essay 6, was presented at a conference on The Logic of Decision and Action at the University of Pittsburgh in March 1966. Comments were made by E. J. Lemmon, H. N. Castañeda, and R. M. Chisholm, and my paper and their comments, along with replies by me, were published in *The Logic of Decision and Action*, edited by Nicholas Rescher, University of Pittsburgh Press, 1967. The sources of the further material at the end of Essay 6 are identified along with the material.

Essay 7, 'Causal Relations', was presented in a symposium with that title at the American Philosophical Association Eastern Division meeting in December 1967; Zeno Vendler commented. Our two papers were published in the *Journal of Philosophy* 64 (1967), and my essay is reprinted by permission of the editors.

The next Essay, number 8, 'The Individuation of Events', was published in *Essays in Honor of Carl G. Hempel*, edited by Nicholas Rescher, D. Reidel 1969, pp. 216–34. Copyright © 1969 by D. Reidel Publishing Company, Dordrecht-Holland. Reprinted by permission of D. Reidel Publishing Company. David Kaplan commented on an earlier draft read at a colloquium at the University of California at Irvine in April 1967. Some of his wisdom is incorporated in the printed version.

Essays 9 and 10, 'Events as Particulars' and 'Eternal *vs*. Ephemeral Events', were touched off by a symposium on events at the May 1970 meeting of the Western Division of the American Philosophical Association. Roderick Chisholm gave the first paper, 'Events and Propositions', and this paper and Essay 9 were first published in *Noûs* 4 (1970), by Wayne State University, Detroit, Michigan. Presently, Chisholm replied to me in 'States of Affairs Again', *Noûs* 5 (1971), and I came back with Essay 10 in the same volume in *Noûs*.

Essay 11, 'Mental Events', was given as one of a series of lectures by various philosophers at the University of Massachusetts in 1968–9. These lectures were published in *Experience and Theory*, edited by Lawrence Foster and J. W. Swanson, The University of Massachusetts Press and Duckworth, 1970, and are reprinted by permission of the publishers.

Essay 12, 'Psychology as Philosophy', was delivered at a symposium on the Philosophy of Psychology at the University of Kent in 1971. The paper was published, with comments and replies, in *Philosophy of Psychology*, edited by S. C. Brown, The Macmillan

Press and Barnes, Noble, Inc., 1974, and is reprinted by permission of Macmillan, London and Basingstoke, and Barnes & Noble Books, New York.

Essay 13, 'The Material Mind', was written in a café in Vienna and delivered at Section 9, Methodology and Philosophy of Psychological Sciences, at the Fourth International Congress for Logic, Methodology, and Philosophy of Science, Bucharest, 1971. The Proceedings, edited by P. Suppes, L. Henkin, G. C. Moisil, and A. Joja, were published by North-Holland Publishing Company in 1973, and my paper is reprinted by permission of the publishers.

Essay 14, 'Hempel on Explaining Action', was read at a celebration in honour of Professor Carl G. Hempel at Princeton University in November 1975, and published in *Erkenntnis* 10 (1976), pp. 239–53. Copyright © 1976 by D. Reidel Publishing Company, Dordrecht-Holland. Reprinted by permission of D. Reidel Publishing Company.

'Hume's Cognitive Theory of Pride', which is Essay 15, was presented at a symposium during the December 1976 meetings of the Eastern Division of the American Philosophical Association, and published in the *Journal of Philosophy* 73 (1976). It is reprinted by permission of the editors. Annette Baier and Keith Donnellan commented on my paper at the symposium, and Annette Baier subsequently expanded her remarks into a paper, 'Hume's Analysis of Pride', the *Journal of Philosophy* 75 (1978).

Essays 1, 3, 6, 7, 8, 11, and 12 have been previously reprinted in various places and languages.

In memory of Nancy Hirschberg

Introduction

All the essays in this book have been published elsewhere, and each was designed to be more or less free standing. But though composed over a baker's dozen of years, they are unified in theme and general thesis. The theme is the role of causal concepts in the description and explanation of human action. The thesis is that the ordinary notion of cause which enters into scientific or common-sense accounts of non-psychological affairs is essential also to the understanding of what it is to act with a reason, to have a certain intention in acting, to be an agent, to act counter to one's own best judgement, or to act freely. Cause is the cement of the universe; the concept of cause is what holds together our picture of the universe, a picture that would otherwise disintegrate into a diptych of the mental and the physical.

Within the three broad subdivisions I have imposed on the essays, the order of publication provides a reasonably natural organizational scheme. One thing led to another, the solutions of one paper raising the problems of the next. All did not go smoothly, however, as will be apparent to even the most sympathetic reader: later problems often prompted fairly drastic reworking of earlier doctrines. Unity of general thesis comes, in these pages, with considerable diachronic inconsistency.

No attempt has been made to conceal the discrepancies between early and later views. Some inadvertent blunders and stylistic uglinesses have been eliminated, and redundance reduced. Redundance in plenty remains, but the points that are worked over most are usually ones that gave me trouble, and so there is, I hope, instruction or interest in what may seem, and probably was intended as, mere repetition. Another reason for leaving my first thoughts substantially unchanged is that over the years they have attracted comment and criticism, and it would be mean spirited to try to move the target out of range after the shot has been fired.

Here are some of the connections among the essays.

Essay 1. 'Actions, Reasons, and Causes' was a reaction against a widely accepted doctrine that the explanation of an intentional action in terms of its motives or reasons could not relate reasons and actions as cause and effect. A principal argument was that causal relations are essentially nomological and based on induction while our knowledge that an agent has acted on certain reasons is not usually dependent on induction or knowledge of serious laws. The argument had found influential if brief expression in Wittgenstein's *Blue and Brown Books*, which were widely circulated from the middle thirties onward (though published only in 1958). In Essay 1 I accept the view that teleological explanation of action differs from explanation in the natural sciences in that laws are not essentially involved in the former but hold that both sorts of explanation can, and often must, invoke causal connections.

Essay 2. The explanation of an intentional action in terms of reasons differs from explanation in the natural sciences in another crucial way: the propositional contents of the explaining attitudes and beliefs must bear a certain logical relation to the description under which the action is explained—a description that gives us an intention with which the action was performed. But what is this logical relation? In Essay 2, 'How is Weakness of the Will Possible?', I assume that no answer is acceptable that frustrates a coherent account of moral conflict, weakness of the will, or other forms of intentional, but irrational, action. In Essay 2 I come out against the view, espoused in Essay 1, that the propositional expressions of the reasons for an action are deductively related to the proposition that corresponds to the action as explained by those reasons.

Causal theories of action are challenged by intentional actions that are contrary to the actor's best judgement. For if reasons are causes, it is natural to suppose that the strongest reasons are the strongest causes. I defend the causal view in Essay 2 by arguing that a reason that is causally strongest need not be a reason deemed by the actor to provide the strongest (best) grounds for acting.

Essay 3 on Agency asks what the relation is between an agent and an event that makes the event an action. A number of suggestions are rejected, and several proposals for reducing some cases to others are made: actions that are not intentional are claimed to be intentional under other descriptions, and actions that seem to include consequences of actions are argued to be identical with the causing action. But the central request for an analysis of agency (and hence, by the argument, of action) goes unanswered.

Essay 4. Causal theories have always been vulnerable to the criticism that they cannot give an acceptable account of free action, or of being free to perform an act. In this Essay I try to defuse some arguments that have been taken to show that freedom to act cannot be a causal power. But I also criticize a number of attempts to analyse that power, and conclude that although freedom to act is a causal power, it cannot be analysed or defined, at least without appeal to the notion of intention. Because of the intimate connection between freedom to act and intentional action, the conclusion of this Essay contradicts an optimistic footnote in Essay 1 which looked forward to, but did not provide, a statement of sufficient conditions of intentional (free) action.

Essay 5. When I wrote Essay 1 I believed that of the three main uses of the concept of intention distinguished by Anscombe (acting with an intention, acting intentionally, and intending to act), the first was the most basic. Acting intentionally, I argued in Essay 1, was just acting with some intention. That left intending, which I somehow thought would be simple to understand in terms of the others. I was wrong. When I finally came to work on it, I found it the hardest of the three; contrary to my original view, it came to seem the basic notion on which the others depend; and what progress I made with it partially undermined an important theme in Essay 1—that 'the intention with which the action was done' does not refer to an entity or state of any kind.

Essay 6. 'The Logical Form of Action Sentences', introduces a group of five papers on the semantics of sentences containing verbs of action or of change generally, and of closely related sentences with noun phrases that seem to refer to events or actions.

In Essay 1 I talked glibly of actions under a description or of two descriptions of the same action, but I was uneasy over the fact that most sentences concerned with actions contained no description or other device that could be taken to refer to an event or action. Some philosophers have regarded this excursion into questions of logical form as extraneous to the theory of action, but this is a risky attitude. For it is clear, I think, that many theories of action that at first appear attractive reveal underlying confusions when subjected to careful semantic scrutiny. When I wrote Essay 1, it had not yet occurred to me that a sentence like 'Eve ate the apple' should not be taken to contain a singular reference to an event; it is distinct in logical form from 'Eve's eating of the apple occurred', though the latter does imply the former.

I have appended to Essay 6 replies to various suggestions and criticisms relevant to its thesis.

Essay 7 applies the lesson of Essay 6 to causality. Essay 1 depended in large part on the fact that events can be described in logically independent ways, so that reasons and actions, described as they must be to bring out the intention with which the action was performed, may not be described as they must be to suggest serious causal laws. Yet reasons and actions can for all that be cause and effect. But in Essay 1 I had no definite idea of what I meant by the description of an event, and therefore no idea either what the logical form of singular causal sentences might be.

Essay 8. The doctrine of Essay 6 dictated that certain descriptions of actions or events, if they refer at all, must refer to the same thing. Thus most adverbial modifiers cannot change the reference of an expression they modify, and (as one would expect—though this is challenged by some current theories) substituting coreferring singular terms in the description of an action or event cannot (except in special contexts) change the action or event referred to.

But these cases do not touch more interesting problems about the individuation of events where logic alone does not decide the matter. Essay 8 is largely concerned with further criteria of individuation. In particular, it first introduced in print (as far as I know) a puzzle that has since come in for a good deal of discussion: if *A* kills *B* by poisoning him, it seems hard to deny that there must be a killing that is identical with a poisoning. But a killing implies a death, and the death may occur long after the act that caused it. How then can the poisoning be identical with the killing? Essay 8 proposed an answer that has been much criticized and defended. The answer still seems to me better than the radical alternatives, though in some respects I now think it can be improved.

Essays 9 and 10 defend and elaborate on the event ontology proposed in Essay 6. Both Essays were written in response to papers by Roderick Chisholm.

The next four papers explore the role of laws in the explanation of actions and other psychological phenomena.

Essay 11 argues that although psychological and physical phenomena are causally connected, and this implies that there are strict laws that cover the instancing events, nevertheless there are no strict laws that cover events or states described in psychological terms. In the course of explaining how this can be, a version of the identity theory emerges which I call *anomalous monism*.

Essay 12 further develops the arguments against the possibility of strict psychophysical laws, and stresses the central importance of a normative concept of rationality in the attribution of beliefs, intentions, desires, and other such attitudes.

Essay 13 raises the question how much we can hope to learn about the psychology of thought and action from advances in neurophysiology and reaches conclusions closely related to those of Essays 11 and 12.

Essay 14, like Essay 8, was written to celebrate the intellectual and other virtues of Carl Hempel. The present Essay, 'Hempel on Explaining Action', acknowledges him as an early partisan of the causal view of action, but takes issue with him on the role of empirical laws in explaining actions.

Essay 15 on Hume's theory of pride defends an Humean analysis of certain passions and emotions, and in particular defends Hume's claim that a belief that one has a beautiful house (for example) is a causal condition of being proud that one has a beautiful house. As in Essay 1 and subsequent Essays I argue that it may be a necessary truth that an event or state, described in a certain way, has a certain cause. Whether or not this is consistent with Hume's analysis of causality is another question, and, in my opinion, an open one. In any case, if the view of the nature of 'propositional' pride and other such emotions which I argue for here is right, it shows how pride, like the actions for which it may provide a reason, enters the causal chains that help explain and to some extent justify human patterns of feeling, thought, and behaviour.

I have many to thank, among them and in particular Max Black, Michael Bratman, Joel Feinberg, Paul Grice, Stuart Hampshire, Gilbert Harman, Carl Hempel, Merrill Hintikka, Georg Kreisel, Sue Larson, David Lewis, Harry Lewis, Mary Mothersill, David Nivison, David Pears, Richard Reiss, Richard Rorty, Allison Ryan, David Sachs, J. J. C. Smart, P. F. Strawson, Patrick Suppes, Irving xvi *Introduction*

Thalberg, David Wiggins, and Kathleen Wilkes. Akeel Bilgrami prepared the bibliography of this book and did much to improve the style and content. With Sue Larson, he helped make the index. I could not have done without their aid and encouragement.

Work on several of the Essays was supported by the National Science Foundation or the Guggenheim Foundation. Two of the Essays were written during a year at the Center for Advanced Study in the Behavioral Sciences at Stanford, and one while I was a visiting fellow at All Souls Çollege, Oxford.

My Gavin David Young Lectures, given at the University of Adelaide in 1968, were mainly drawn from drafts of articles printed here. I am grateful to a number of people who commented on those lectures, especially to Jack Smart whose views on the relation of the mental and the physical, while in some respects orthogonal to mine, have much stimulated my own thinking. I am also grateful to him for pointing out to me an astronomical (in subject matter, not size) error.

Most of the ideas in this book were worked out and tested in discussion with those whom I was at one time supposed to be teaching. Daniel Bennett did much to get me started on the subject. His dissertation, in theory written under my direction, opened my eyes to what was going on in Oxford in the middle fifties, and I was further influenced when, as colleagues at Stanford University, we gave a seminar together on action theory.

John Wallace and I talked endlessly about problems in the philosophy of language, first when he was a student at Stanford, then during a year in Athens, Corfu, and Gründelwald, and subsequently when we were colleagues at Princeton University and the Rockefeller University. The key ideas of Essays 6 and 7 sprang from these talks.

In 1968 Nancy Hirschberg invited me to give a talk to the Psychology Department at the University of Illinois in Champaign; Essay 11 was one result. In the composition of the last five Essays I was guided by her knowledge of psychology and candid advice. Her enthusiasm, gaiety, and affection made the work fun and the fun wonderful.

INTENTION AND ACTION

1 *Actions, Reasons, and Causes*

What is the relation between a reason and an action when the reason explains the action by giving the agent's reason for doing what he did? We may call such explanations *rationalizations*, and say that the reason *rationalizes* the action.

In this paper I want to defend the ancient—and common-sense—position that rationalization is a species of causal explanation. The defence no doubt requires some redeployment, but it does not seem necessary to abandon the position, as has been urged by many recent writers.[1]

I

A reason rationalizes an action only if it leads us to see something the agent saw, or thought he saw, in his action—some feature, consequence, or aspect of the action the agent wanted, desired, prized, held dear, thought dutiful, beneficial, obligatory, or agreeable. We cannot explain why someone did what he did simply by saying the particular action appealed to him; we must indicate what it was about the action that appealed. Whenever someone does something for a reason, therefore, he can be characterized as (*a*) having some sort of pro attitude toward actions of a certain kind, and (*b*) believing (or knowing, perceiving, noticing, remembering)

[1] Some examples: Gilbert Ryle, *The Concept of Mind*, G. E. M. Anscombe, *Intention*, Stuart Hampshire, *Thought and Action*, H. L. A. Hart and A. M. Honoré, *Causation in the Law*, William Dray, *Laws and Explanation in History*, and most of the books in the series edited by R. F. Holland, *Studies in Philosophical Psychology*, including Anthony Kenny, *Action, Emotion and Will*, and A. I. Melden, *Free Action*. Page references in parentheses are to these works.

that his action is of that kind. Under (*a*) are to be included desires, wantings, urges, promptings, and a great variety of moral views, aesthetic principles, economic prejudices, social conventions, and public and private goals and values in so far as these can be interpreted as attitudes of an agent directed toward actions of a certain kind. The word 'attitude' does yeoman service here, for it must cover not only permanent character traits that show themselves in a lifetime of behaviour, like love of children or a taste for loud company, but also the most passing fancy that prompts a unique action, like a sudden desire to touch a woman's elbow. In general, pro attitudes must not be taken for convictions, however temporary, that every action of a certain kind ought to be performed, is worth performing, or is, all things considered, desirable. On the contrary, a man may all his life have a yen, say, to drink a can of paint, without ever, even at the moment he yields, believing it would be worth doing.

Giving the reason why an agent did something is often a matter of naming the pro attitude (*a*) or the related belief (*b*) or both; let me call this pair the *primary reason* why the agent performed the action. Now it is possible both to reformulate the claim that rationalizations are causal explanations and to give structure to the argument by stating two theses about primary reasons:

1. In order to understand how a reason of any kind rationalizes an action it is necessary and sufficient that we see, at least in essential outline, how to construct a primary reason.
2. The primary reason for an action is its cause.

I shall argue for these points in turn.

II

I flip the switch, turn on the light, and illuminate the room. Unbeknownst to me I also alert a prowler to the fact that I am home. Here I need not have done four things, but only one, of which four descriptions have been given.[2] I flipped the switch because I wanted

---

[2] We might not call my unintentional alerting of the prowler an action, but it should not be inferred from this that alerting the prowler is therefore something different from flipping the switch, say just its consequence. Actions, performances, and events not involving intention are alike in that they are often referred to or defined partly in terms of some terminal stage, outcome, or consequence.

to turn on the light and by saying I wanted to turn on the light I explain (give my reason for, rationalize) the flipping. But I do not, by giving this reason, rationalize my alerting of the prowler nor my illuminating of the room. Since reasons may rationalize what someone does when it is described in one way and not when it is described in another, we cannot treat what was done simply as a term in sentences like 'My reason for flipping the switch was that I wanted to turn on the light'; otherwise we would be forced to conclude, from the fact that flipping the switch was identical with alerting the prowler, that my reason for alerting the prowler was that I wanted to turn on the light. Let us mark this quasi-intensional[3] character of action descriptions in rationalizations by stating a bit more precisely a necessary condition for primary reasons:

C1. *R* is a primary reason why an agent performed the action *A* under the description *d* only if *R* consists of a pro attitude of the agent towards actions with a certain property, and a belief of the agent that *A*, under the description *d*, has that property.

How can my wanting to turn on the light be (part of) a primary reason, since it appears to lack the required element of generality? We may be taken in by the verbal parallel between 'I turned on the light' and 'I wanted to turn on the light'. The first clearly refers to a particular event, so we conclude that the second has this same event as its object. Of course it is obvious that the event of my turning on the light can't be referred to in the same way by both sentences since the existence of the event is required by the truth of 'I turned

---

The word 'action' does not very often occur in ordinary speech, and when it does it is usually reserved for fairly portentous occasions. I follow a useful philosophical practice in calling anything an agent does intentionally an action, including intentional omissions. What is really needed is some suitably generic term to bridge the following gap: suppose '*A*' is a description of an action, '*B*' is a description of something done voluntarily, though not intentionally, and '*C*' is a description of something done involuntarily and unintentionally; finally, suppose *A* = *B* = *C*. Then *A*, *B*, and *C* are the same—what? 'Action', 'event', 'thing done', each have, at least in some contexts, a strange ring when coupled with the wrong sort of description. Only the question, 'Why did you (he) do *A*?' has the true generality required. Obviously, the problem is greatly aggravated if we assume, as Melden does, that an action ('raising one's arm') can be identical with a bodily movement ('one's arm going up').

[3] 'Quasi-intentional' because, besides its intentional aspect, the description of the action must also refer in rationalizations; otherwise it could be true that an action was done for a certain reason and yet the action not have been performed. Compare 'the author of *Waverley*' in 'George IV knew the author of Waverley wrote *Waverley*'. This semantical feature of action descriptions is discussed further in Essays 3 and 6.

on the light' but not by the truth of 'I wanted to turn on the light'. If the reference were the same in both cases, the second sentence would entail the first; but in fact the sentences are logically independent. What is less obvious, at least until we attend to it, is that the event whose occurrence makes 'I turned on the light' true cannot be called the object, however intentional, of 'I wanted to turn on the light'. If I turned on the light, then I must have done it at a precise moment, in a particular way—every detail is fixed. But it makes no sense to demand that my want be directed to an action performed at any one moment or done in some unique manner. Any one of an indefinitely large number of actions would satisfy the want and can be considered equally eligible as its object. Wants and desires often are trained on physical objects. However, 'I want that gold watch in the window' is not a primary reason and explains why I went into the store only because it suggests a primary reason—for example, that I wanted to buy the watch.

Because 'I wanted to turn on the light' and 'I turned on the light' are logically independent, the first can be used to give a reason why the second is true. Such a reason gives minimal information: it implies that the action was intentional, and wanting tends to exclude some other pro attitudes, such as a sense of duty or obligation. But the exclusion depends very much on the action and the context of explanation. Wanting seems pallid beside lusting, but it would be odd to deny that someone who lusted after a woman or a cup of coffee wanted her or it. It is not unnatural, in fact, to treat wanting as a genus including all pro attitudes as species. When we do this and when we know some action is intentional, it is easy to answer the question, 'Why did you do it?' with, 'For no reason', meaning not that there is no reason but that there is no *further* reason, no reason that cannot be inferred from the fact that the action was done intentionally; no reason, in other words, besides wanting to do it. This last point is not essential to the present argument, but it is of interest because it defends the possibility of defining an intentional action as one done for a reason.

A primary reason consists of a belief and an attitude, but it is generally otiose to mention both. If you tell me you are easing the jib because you think that will stop the main from backing, I don't need to be told that you want to stop the main from backing; and if you say you are biting your thumb at me because you want to insult me, there is no point in adding that you think that by biting your thumb at me you will insult me. Similarly, many explanations of actions in terms of reasons that are not primary do not require mention of the primary reason to complete the story. If I say I am pulling weeds because I want a beautiful lawn, it would be fatuous to eke out the account with, 'And so I see something desirable in any action that does, or has a good chance of, making the lawn beautiful'. Why insist that there is any *step*, logical or psychological, in the transfer of desire from an end that is not an action to the actions one conceives as means? It serves the argument as well that the desired end explains the action only if what are believed by the agent to be means are desired.

Fortunately, it is not necessary to classify and analyse the many varieties of emotions, sentiments, moods, motives, passions, and hungers whose mention may answer the question, 'Why did you do it?' in order to see how, when such mention rationalizes the action, a primary reason is involved. Claustrophobia gives a man's reason for leaving a cocktail party because we. know people want to avoid, escape from, be safe from, put distance between themselves and what they fear. Jealousy is the motive in a poisoning because, among other things, the poisoner believes his action will harm his rival, remove the cause of his agony, or redress an injustice, and these are the sorts of things a jealous man wants to do. When we learn that a man cheated his son out of greed, we do not necessarily know what the primary reason was, but we know there was one, and its general nature. Ryle analyses 'he boasted from vanity' into 'he boasted on meeting the stranger and his doing so satisfies the lawlike proposition that whenever he finds a chance of securing the admiration and envy of others, he does whatever he thinks will produce this admiration and envy' (89). This analysis is often, and perhaps justly, criticized on the ground that a man may boast from vanity just once. But if Ryle's boaster did what he did from vanity, then something entailed by Ryle's analysis is true: the boaster wanted to secure the admiration and envy of others, and he believed that his action would produce this admiration and envy; true or false, Ryle's analysis does not dispense with primary reasons, but depends upon them.

To know a primary reason why someone acted as he did is to know an intention with which the action was done. If I turn left at the fork because I want to get to Katmandu, my intention in turning left is to get to Katmandu. But to know the intention is not necessarily to know the primary reason in full detail. If James goes to church with the intention of pleasing his mother, then he must have some pro attitude toward pleasing his mother, but it needs more information to tell whether his reason is that he enjoys pleasing his mother, or thinks it right, his duty, or an obligation. The expression 'the intention with which James went to church' has the outward form of a description, but in fact it is syncategorematic and cannot be taken to refer to an entity, state, disposition, or event. Its function in context is to generate new descriptions of actions in terms of their reasons; thus 'James went to church with the intention of pleasing his mother' yields a new, and fuller, description of the action described in 'James went to church'. Essentially the same process goes on when I answer the question, 'Why are you bobbing around that way?' with, 'I'm knitting, weaving, exercising, sculling, cuddling, training fleas'.

Straight description of an intended result often explains an action better than stating that the result was intended or desired. 'It will soothe your nerves' explains why I pour you a shot as efficiently as 'I want to do something to soothe your nerves', since the first in the context of explanation implies the second; but the first does better, because, if it is true, the facts will justify my choice of action. Because justifying and explaining an action so often go hand in hand, we frequently indicate the primary reason for an action by making a claim which, if true, would also verify, vindicate, or support the relevant belief or attitude of the agent. 'I knew I ought to return it', 'The paper said it was going to snow', 'You stepped on *my* toes', all, in appropriate reason-giving contexts, perform this familiar dual function.

The justifying role of a reason, given this interpretation, depends upon the explanatory role, but the converse does not hold. Your stepping on my toes neither explains nor justifies my stepping on your toes unless I believe you stepped on my toes, but the belief alone, true or false, explains my action.

III

In the light of a primary reason, an action is revealed as coherent with certain traits, long- or short-termed, characteristic or not, of the agent, and the agent is shown in his role of Rational Animal.

Corresponding to the belief and attitude of a primary reason for an action, we can always construct (with a little ingenuity) the premises of a syllogism from which it follows that the action has some (as Anscombe calls it) 'desirability characteristic'.[4] Thus there is a certain irreducible—though somewhat anaemic—sense in which every rationalization justifies: from the agent's point of view there was, when he acted, something to be said for the action.

Noting that nonteleological causal explanations do not display the element of justification provided by reasons, some philosophers have concluded that the concept of cause that applies elsewhere cannot apply to the relation between reasons and actions, and that the pattern of justification provides, in the case of reasons, the required explanation. But suppose we grant that reasons alone justify actions in the course of explaining them; it does not follow that the explanation is not also—and necessarily—causal. Indeed our first condition for primary reasons (C1) is designed to help set rationalizations apart from other sorts of explanation. If rationalization is, as I want to argue, a species of causal explanation, then justification, in the sense given by C1, is at least one differentiating property. How about the other claim: that justifying is a kind of explaining, so that the ordinary notion of cause need not be brought in? Here it is necessary to decide what is being included under justification. It could be taken to cover only what is called for by C1: that the agent have certain beliefs and attitudes in the light of which the action is reasonable. But then something essential has certainly been left out, for a person can have a reason for an action, and perform the action, and yet this reason not be the reason why he did it. Central to the relation between a reason and an action it explains is the idea that the agent performed the action *because* he had the reason. Of course, we can include this idea too in justification; but then the notion of justification becomes as dark as the notion of reason until we can account for the force of that 'because'.

When we ask why someone acted as he did, we want to be

---

[4] Anscombe denies that the practical syllogism is deductive. This she does partly because she thinks of the practical syllogism, as Aristotle does, as corresponding to a piece of practical reasoning (whereas for me it is only part of the analysis of the concept of a reason with which someone acted), and therefore she is bound, again following Aristotle, to think of the conclusion of a practical syllogism as corresponding to a judgement, not merely that the action has a desirable characteristic, but that the action is desirable (reasonable, worth doing, etc.). Practical reasoning is discussed further in Essay 2.

provided with an interpretation. His behaviour seems strange, alien, outré, pointless, out of character, disconnected; or perhaps we cannot even recognize an action in it. When we learn his reason, we have an interpretation, a new description of what he did, which fits it into a familiar picture. The picture includes some of the agent's beliefs and attitudes; perhaps also goals, ends, principles, general character traits, virtues or vices. Beyond this, the redescription of an action afforded by a reason may place the action in a wider social, economic, linguistic, or evaluative context. To learn, through learning the reason, that the agent conceived his action as a lie, a repayment of a debt, an insult, the fulfilment of an avuncular obligation, or a knight's gambit is to grasp the point of the action in its setting of rules, practices, conventions, and expectations.

Remarks like these, inspired by the later Wittgenstein, have been elaborated with subtlety and insight by a number of philosophers. And there is no denying that this is true: when we explain an action, by giving the reason, we do redescribe the action; redescribing the action gives the action a place in a pattern, and in this way the action is explained. Here it is tempting to draw two conclusions that do not follow. First, we can't infer, from the fact that giving reasons merely redescribes the action and that causes are separate from effects, that therefore reasons are not causes. Reasons, being beliefs and attitudes, are certainly not identical with actions; but, more important, events are often redescribed in terms of their causes. (Suppose someone was injured. We could redescribe this event 'in terms of a cause' by saying he was burned.) Second, it is an error to think that, because placing the action in a larger pattern explains it, therefore we now understand the sort of explanation involved. Talk of patterns and contexts does not answer the question of how reasons explain actions, since the relevant pattern or context contains both reason and action. One way we can explain an event is by placing it in the context of its cause; cause and effect form the sort of pattern that explains the effect, in a sense of 'explain' that we understand as well as any. If reason and action illustrate a different pattern of explanation, that pattern must be identified.

Let me urge the point in connection with an example of Melden's. A man driving an automobile raises his arm in order to signal. His intention, to signal, explains his action, raising his arm, by redescribing it as signalling. What is the pattern that explains the action? Is it the familiar pattern of an action done for a reason? Then it does indeed explain the action, but only because it assumes the relation of reason and action that we want to analyse. Or is the pattern rather this: the man is driving, he is approaching a turn; he knows he ought to signal; he knows how to signal, by raising his arm. And now, in this context, he raises his arm. Perhaps, as Melden suggests, if all this happens, he does signal. And the explanation would then be this; if, under these conditions, a man raises his arm, then he signals. The difficulty is, of course, that this explanation does not touch the question of why he raised his arm. He had a reason to raise his arm, but this has not been shown to be the reason why he did it. If the description 'signalling' explains his action by giving his reason, then the signalling must be intentional; but, on the account just given, it may not be.

If, as Melden claims, causal explanations are 'wholly irrelevant to the understanding we seek' of human action (184) then we are without an analysis of the 'because' in 'He did it because ...', where we go on to name a reason. Hampshire remarks, of the relation between reasons and action, 'In philosophy one ought surely to find this ... connection altogether mysterious' (166). Hampshire rejects Aristotle's attempt to solve the mystery by introducing the concept of wanting as a causal factor, on the grounds that the resulting theory is too clear and definite to fit all cases and that, 'There is still no compelling ground for insisting that the word "want" *must* enter into every full statement of reasons for acting' (168). I agree that the concept of wanting is too narrow, but I have argued that, at least in a vast number of typical cases, some pro attitude must be assumed to be present if a statement of an agent's reasons in acting is to be intelligible. Hampshire does not see how Aristotle's scheme can be appraised as true or false, 'for it is not clear what could be the basis of assessment, or what kind of evidence could be decisive' (167). But I would urge that, failing a satisfactory alternative, the best argument for a scheme like Aristotle's is that it alone promises to give an account of the 'mysterious connection' between reasons and actions.

IV

In order to turn the first 'and' to 'because' in 'He exercised *and* he wanted to reduce and thought exercise would do it', we must, as the basic move,[5] augment condition C1 with:

C2. A primary reason for an action is its cause.

The considerations in favour of C2 are by now, I hope, obvious; in the remainder of this paper I wish to defend C2 against various lines of attack and, in the process, to clarify the notion of causal explanation involved.

A. The first line of attack is this. Primary reasons consist of attitudes and beliefs, which are states or dispositions, not events; therefore they cannot be causes.

It is easy to reply that states, dispositions, and conditions are frequently named as the causes of events: the bridge collapsed because of a structural defect; the plane crashed on takeoff because the air temperature was abnormally high; the plate broke because it had a crack. This reply does not, however, meet a closely related point. Mention of a causal condition for an event gives a cause only on the assumption that there was also a preceding event. But what is the preceding event that causes an action?

In many cases it is not difficult at all to find events very closely associated with the primary reason. States and dispositions are not events, but the onslaught of a state or disposition is. A desire to hurt your feelings may spring up at the moment you anger me; I may start wanting to eat a melon just when I see one; and beliefs may begin at the moment we notice, perceive, learn, or remember something. Those who have argued that there are no mental events to qualify as causes of actions have often missed the obvious because they have insisted that a mental event be observed or noticed (rather than an observing or a noticing) or that it be like a stab, a qualm, a prick or a quiver, a mysterious prod of conscience or act of the will. Melden, in discussing the driver who signals a turn by raising his arm, challenges those who want to explain actions causally to identify 'an event which is common and peculiar to all such cases' (87), perhaps a motive or an intention, anyway 'some particular feeling or experience' (95). But of course there is a mental event; at some moment the driver noticed (or thought he noticed) his turn coming up, and that is the moment he signalled. During any continu-

---

[5] I say 'as the basic move' to cancel any suggestion that C1 and C2 are jointly *sufficient* to define the relation of reasons to the actions they explain. For discussion of this point, see the Introduction and Essay 4.

ing activity, like driving, or elaborate performance, like swimming the Hellespont, there are more or less fixed purposes, standards, desires, and habits that give direction and form to the entire enterprise, and there is the continuing input of information about what we are doing, about changes in the environment, in terms of which we regulate and adjust our actions. To dignify a driver's awareness that his turn has come by calling it an experience, or even a feeling, is no doubt exaggerated, but whether it deserves a name or not, it had better be the reason why he raises his arm. In this case, and typically, there may not be anything we would call a motive, but if we mention such a general purpose as wanting to get to one's destination safely, it is clear that the motive is not an event. The intention with which the driver raises his arm is also not an event, for it is no thing at all, neither event, attitude, disposition, nor object. Finally, Melden asks the causal theorist to find an event that is common and peculiar to all cases where a man intentionally raises his arm, and this, it must be admitted, cannot be produced. But then neither can a common and unique cause of bridge failures, plane crashes, or plate breakings be produced.

The signalling driver can answer the question, 'Why did you raise your arm when you did?', and from the answer we learn the event that caused the action. But can an actor always answer such a question? Sometimes the answer will mention a mental event that does not give a reason: 'Finally I made up my mind.' However, there also seem to be cases of intentional action where we cannot explain at all why we acted when we did. In such cases, explanation in terms of primary reasons parallels the explanation of the collapse of the bridge from a structural defect: we are ignorant of the event or sequence of events that led up to (caused) the collapse, but we are sure there was such an event or sequence of events.

B. According to Melden, a cause must be 'logically distinct from the alleged effect' (52); but a reason for an action is not logically distinct from the action; therefore, reasons are not causes of actions.[6]

One possible form of this argument has already been suggested. Since a reason makes an action intelligible by redescribing it, we do

---

[6] This argument can be found in one or more versions, in Kenny, Hampshire, and Melden, as well as in P. Winch, *The Idea of a Social Science*, and R. S. Peters, *The Concept of Motivation*. In one of its forms, the argument was of course inspired by Ryle's treatment of motives in *The Concept of Mind*.

not have two events, but only one under different descriptions. Causal relations, however, demand distinct events.

Someone might be tempted into the mistake of thinking that my flipping of the switch caused my turning on of the light (in fact it caused the light to go on). But it does not follow that it is a mistake to take, 'My reason for flipping the switch was that I wanted to turn on the light' as entailing, in part, 'I flipped the switch, and this action is further describable as having been caused by wanting to turn on the light'. To describe an event in terms of its cause is not to confuse the event with its cause, nor does explanation by redescription exclude causal explanation.

The example serves also to refute the claim that we cannot describe the action without using words that link it to the alleged cause. Here the action is to be explained under the description: 'my flipping the switch', and the alleged cause is 'my wanting to turn on the light'. What relevant logical relation is supposed to hold between these phrases? It seems more plausible to urge a logical link between 'my turning on the light' and 'my wanting to turn on the light', but even here the link turns out, on inspection, to be grammatical rather than logical.

In any case there is something very odd in the idea that causal relations are empirical rather than logical. What can this mean? Surely not that every true causal statement is empirical. For suppose 'A caused B' is true. Then the cause of $B = A$; so substituting, we have 'The cause of B caused B', which is analytic. The truth of a causal statement depends on *what* events are described; its status as analytic or synthetic depends on *how* the events are described. Still, it may be maintained that a reason rationalizes an action only when the descriptions are appropriately fixed, and the appropriate descriptions are not logically independent.

Suppose that to say a man wanted to turn on the light *meant* that he would perform any action he believed would accomplish his end. Then the statement of his primary reason for flipping the switch would entail that he flipped the switch—'straightway he acts', as Aristotle says. In this case there would certainly be a logical connection between reason and action, the same sort of connection as that between, 'It's water-soluble and was placed in water' and 'It dissolved'. Since the implication runs from description of cause to description of effect but not conversely, naming the cause still gives information. And, though the point is often overlooked, 'Placing it in water caused it to dissolve' does not entail 'It's water-soluble'; so the latter has additional explanatory force. Nevertheless, the explanation would be far more interesting if, in place of solubility, with its obvious definitional connection with the event to be explained, we could refer to some property, say a particular crystalline structure, whose connection with dissolution in water was known only through experiment. Now it is clear why primary reasons like desires and wants do not explain actions in the relatively trivial way solubility explains dissolvings. Solubility, we are assuming, is a pure disposition property: it is defined in terms of a single test. But desires cannot be defined in terms of the actions they may rationalize, even though the relation between desire and action is not simply empirical; there are other, equally essential criteria for desires—their expression in feelings and in actions that they do not rationalize, for example. The person who has a desire (or want or belief) does not normally need criteria at all—he generally knows, even in the absence of any clues available to others, what he wants, desires, and believes. These logical features of primary reasons show that it is not just lack of ingenuity that keeps us from defining them as dispositions to act for these reasons.

C. According to Hume, 'we may define a cause to be an object, followed by another, and where all the objects similar to the first are followed by objects similar to the second'. But, Hart and Honoré claim, 'The statement that one person did something because, for example, another threatened him, carries no implication or covert assertion that if the circumstances were repeated the same action would follow' (52). Hart and Honoré allow that Hume is right in saying that ordinary singular causal statements imply generalizations, but wrong for this very reason in supposing that motives and desires are ordinary causes of actions. In brief, laws are involved essentially in ordinary causal explanations, but not in rationalizations.

It is common to try to meet this argument by suggesting that we do have rough laws connecting reasons and actions, and these can, in theory, be improved. True, threatened people do not always respond in the same way; but we may distinguish between threats and also between agents, in terms of their beliefs and attitudes.

The suggestion is delusive, however, because generalizations connecting reasons and actions are not—and cannot be sharpened into—the kind of law on the basis of which accurate predictions can reliably be made. If we reflect on the way in which reasons determine choice, decision, and behaviour, it is easy to see why this is so. What emerges, in the *ex post facto* atmosphere of explanation and justification, as *the* reason frequently was, to the agent at the time of action, one consideration among many, *a* reason. Any serious theory for predicting action on the basis of reasons must find a way of evaluating the relative force of various desires and beliefs in the matrix of decision; it cannot take as its starting point the refinement of what is to be expected from a single desire. The practical syllogism exhausts its role in displaying an action as falling under one reason; so it cannot be subtilized into a reconstruction of practical reasoning, which involves the weighing of competing reasons. The practical syllogism provides a model neither for a predictive science of action nor for a normative account of evaluative reasoning.

Ignorance of competent predictive laws does not inhibit valid causal explanation, or few causal explanations could be made. I am certain the window broke because it was struck by a rock—I saw it all happen; but I am not (is anyone?) in command of laws on the basis of which I can predict what blows will break which windows. A generalization like, 'Windows are fragile, and fragile things tend to break when struck hard enough, other conditions being right' is not a predictive law in the rough—the predictive law, if we had it, would be quantitative and would use very different concepts. The generalization, like our generalizations about behaviour, serves a different function: it provides evidence for the existence of a causal law covering the case at hand.[7]

We are usually far more certain of a singular causal connection than we are of any causal law governing the case; does this show that Hume was wrong in claiming that singular causal statements entail laws? Not necessarily, for Hume's claim, as quoted above, is ambiguous. It may mean that '*A* caused *B*' entails some particular law involving the predicates used in the descriptions '*A*' and '*B*', or it may mean that '*A* caused *B*' entails that there exists a causal law instantiated by some true descriptions of *A* and *B*.[8] Obviously, both

---

[7] Essays 11, 12, and 13 discuss the issues of this paragraph and the one before it.

[8] We could roughly characterize the analysis of singular causal statements hinted at here as follows: '*A* caused *B*' is true if and only if there are descriptions of *A* and *B* such that the sentence obtained by putting these descriptions for '*A*' and '*B*' in '*A* caused *B*' follows from a true causal law. This analysis is saved from triviality by the fact that not all true generalizations are causal laws; causal laws are distinguished (though of course this is no analysis) by the fact that they are inductively confirmed by their instances and by the fact that they support counterfactual and subjunctive singular causal statements. There is more on causality in Essay 7.

versions of Hume's doctrine give a sense to the claim that singular causal statements entail laws, and both sustain the view that causal explanations 'involve laws'. But the second version is far weaker, in that no particular law is entailed by a singular causal claim, and a singular causal claim can be defended, if it needs defence, without defending any law. Only the second version of Hume's doctrine can be made to fit with most causal explanations; it suits rationalizations equally well.

The most primitive explanation of an event gives its cause; more elaborate explanations may tell more of the story, or defend the singular causal claim by producing a relevant law or by giving reasons for believing such exists. But it is an error to think no explanation has been given until a law has been produced. Linked with these errors is the idea that singular causal statements necessarily indicate, by the concepts they employ, the concepts that will occur in the entailed law. Suppose a hurricane, which is reported on page 5 of Tuesday's *Times*, causes a catastrophe, which is reported on page 13 of Wednesday's *Tribune*. Then the event reported on page 5 of Tuesday's *Times* caused the event reported on page 13 of Wednesday's *Tribune*. Should we look for a law relating events of these *kinds*? It is only slightly less ridiculous to look for a law relating hurricanes and catastrophes. The laws needed to predict the catastrophe with precision would, of course, have no use for concepts like hurricane and catastrophe. The trouble with predicting the weather is that the descriptions under which events interest us—'a cool, cloudy day with rain in the afternoon'—have only remote connections with the concepts employed by the more precise known laws.

The laws whose existence is required if reasons are causes of actions do not, we may be sure, deal in the concepts in which rationalizations must deal. If the causes of a class of events (actions) fall in a certain class (reasons) and there is a law to back each singular causal statement, it does not follow that there is any law connecting events classified as reasons with events classified as actions—the classifications may even be neurological, chemical, or physical.

D. It is said that the kind of knowledge one has of one's own reasons in acting is not compatible with the existence of a causal relation between reasons and actions: a person knows his own intentions in acting infallibly, without induction or observation, and no ordinary causal relation can be known in this way. No doubt our knowledge of our own intentions in acting will show many of the oddities peculiar to first-person knowledge of one's own pains, beliefs, desires, and so on; the only question is whether these oddities prove that reasons do not cause, in any ordinary sense at least, the actions that they rationalize.

You may easily be wrong about the truth of a statement of the form 'I am poisoning Charles because I want to save him pain', because you may be wrong about whether you are poisoning Charles—you may yourself be drinking the poisoned cup by mistake. But it also seems that you may err about your reasons, particularly when you have two reasons for an action, one of which pleases you and one which does not. For example, you do want to save Charles pain; you also want him out of the way. You may be wrong about which motive made you do it.

The fact that you may be wrong does not show that in general it makes sense to ask you how you know what your reasons were or to ask for your evidence. Though you may, on rare occasions, accept public or private evidence as showing you are wrong about your reasons, you usually have no evidence and make no observations. Then your knowledge of your own reasons for your actions is not generally inductive, for where there is induction, there is evidence. Does this show the knowledge is not causal? I cannot see that it does.

Causal laws differ from true but nonlawlike generalizations in that their instances confirm them; induction is, therefore, certainly a good way to learn the truth of a law. It does not follow that it is the only way to learn the truth of a law. In any case, in order to know that a singular causal statement is true, it is not necessary to know the truth of a law; it is necessary only to know that some law covering the events at hand exists. And it is far from evident that induction, and induction alone, yields the knowledge that a causal law satisfying certain conditions exists. Or, to put it differently, one case is often enough, as Hume admitted, to persuade us that a law exists, and this amounts to saying that we are persuaded without direct inductive evidence, that a causal relation exists.

E. Finally I should like to say something about a certain uneasiness some philosophers feel in speaking of causes of actions at all. Melden, for example, says that actions are often identical with bodily movements, and that bodily movements have causes; yet he denies that the causes are causes of the actions. This is, I think, a contradiction. He is led to it by the following sort of consideration: 'It is futile to attempt to explain conduct through the causal efficacy of desire—all *that* can explain is further happenings, not actions performed by agents. The agent confronting the causal nexus in which such happenings occur is a helpless victim of all that occurs in and to him' (128, 129). Unless I am mistaken, this argument, if it were valid, would show that actions cannot have causes at all. I shall not point out the obvious difficulties in removing actions from the realm of causality entirely. But perhaps it is worth trying to uncover the source of the trouble. Why on earth should a cause turn an action into a mere happening and a person into a helpless victim? Is it because we tend to assume, at least in the arena of action, that a cause demands a causer, agency an agent? So we press the question; if my action is caused, what caused it? If I did, then there is the absurdity of infinite regress; if I did not, I am a victim. But of course the alternatives are not exhaustive. Some causes have no agents. Among these agentless causes are the states and changes of state in persons which, because they are reasons as well as causes, constitute certain events free and intentional actions.

2 How is Weakness of the Will Possible?

An agent's will is weak if he acts, and acts intentionally, counter to his own best judgement; in such cases we sometimes say he lacks the willpower to do what he knows, or at any rate believes, would, everything considered, be better. It will be convenient to call actions of this kind incontinent actions, or to say that in doing them the agent acts incontinently. In using this terminology I depart from tradition, at least in making the class of incontinent actions larger than usual. But it is the larger class I want to discuss, and I believe it includes all of the actions some philosophers have called incontinent, and some of the actions many philosophers have called incontinent.

Let me explain how my conception of incontinence is more general than some others. It is often made a condition of an incontinent action that it be performed despite the agent's knowledge that another course of action is better. I count such actions incontinent, but the puzzle I shall discuss depends only on the attitude or belief of the agent, so it would restrict the field to no purpose to insist on knowledge. Knowledge also has an unneeded, and hence unwanted, flavour of the cognitive; my subject concerns evaluative judgements, whether they are analysed cognitively, prescriptively, or otherwise. So even the concept of belief is perhaps too special, and I shall speak of what the agent judges or holds.

If a man holds some course of action to be the best one, everything considered, or the right one, or the thing he ought to do, and yet does something else, he acts incontinently. But I would also say he acts incontinently provided he holds some available course of action to be better on the whole than the one he takes; or that, as between some other course of action which he believes open to him and the action he performs, he judges that he ought to perform the other. In other words, comparative judgements suffice for incontinence. We may now characterize an action that reveals weakness of the will or incontinence:

> D. In doing x an agent acts incontinently if and only if: (*a*) the agent does x intentionally; (*b*) the agent believes there is an alternative action y open to him; and (*c*) the agent judges that, all things considered, it would be better to do y than to do x.[1]

There seem to be incontinent actions in this sense. The difficulty is that their existence challenges another doctrine that has an air of self-evidence: that, in so far as a person acts intentionally he acts, as Aquinas puts it, in the light of some imagined good. This view does not, as it stands, directly contradict the claim that there are incontinent actions. But it is hard to deny that the considerations that recommend this view recommend also a relativized version: in so far as a person acts intentionally he acts in the light of what he imagines (judges) to be the better.

It will be useful to spell out this claim in the form of two principles. The first expresses the natural assumption about the relation between wanting or desiring something, and action. 'The primitive sign of wanting is trying to get', says Anscombe in *Intention*.[2] Hampshire comes closer to exactly what I need when he writes, in *Freedom of the Individual*,[3] that 'A wants to do X' is equivalent to 'other things being equal, he would do X, if he could'. Here I take (possibly contrary to Hampshire's intent) 'other things being equal' to mean, or anyway to allow, the interpretation, 'provided there

---

[1] In a useful article, G. Santas gives this account of incontinence: 'In a case of weakness a man does something that he knows or believes he should (ought) not do, or fails to do something that he knows or believes he should do, when the occasion and the opportunity for acting or refraining is present, and when it is in his power, in some significant sense, to act in accordance with his knowledge or belief.' ('Plato's *Protagoras* and Explanations of Weakness', 3.) Most of the differences between this description and mine are due to my deliberate deviation from the tradition. But there seem to me to be the two minor errors in Santas's account. First, weakness of the will does not require that the alternative action actually be available, only that the agent think it is. What Santas is after, and correctly, is that the agent acts freely; but for this it is not necessary that the alternative the agent thinks better (or that he ought to do) be open to him. On the other hand (and this is the second point), Santas's criteria are not sufficient to guarantee that the agent acts intentionally, and this is, I think, essential to incontinence.

[2] G. E. M. Anscombe, *Intention*, 67.

[3] S. Hampshire, *Freedom of the Individual*, 36.

is not something he wants more'. Given this interpretation, Hampshire's principle could perhaps be put:

P1. If an agent wants to do x more than he wants to do y and he believes himself free to do either x or y, then he will intentionally do x if he does either x or y intentionally.

The second principle connects judgements of what it is better to do with motivation or wanting:

P2. If an agent judges that it would be better to do x than to do y, then he wants to do x more than he wants to do y.

P1 and P2 together obviously entail that if an agent judges that it would be better for him to do x than to do y, and he believes himself to be free to do either x or y, then he will intentionally do x if he does either x or y intentionally. This conclusion, I suggest, appears to show that it is false that:

P3. There are incontinent actions.

Someone who is convinced that P1–P3 form an inconsistent triad, but who finds only one or two of the principles really persuasive, will have no difficulty deciding what to say. But for someone (like myself) to whom the principles expressed by P1–P3 seem self-evident, the problem posed by the apparent contradiction is acute enough to be called a paradox. I cannot agree with Lemmon when he writes, in an otherwise admirable article, 'Perhaps akrasia is one of the best examples of a pseudo-problem in philosophical literature: in view of its existence, if you find it a problem you have already made a philosophical mistake.'[4] If your assumptions lead to a contradiction, no doubt you have made a mistake, but since you can know you have made a mistake without knowing what the mistake is, your problem may be real.

The attempted solutions with which I am familiar to the problem created by the initial plausibility of P1–P3 assume that P1–P3 do really contradict one another. These attempts naturally end by giving up one or another of the principles. I am not very happy about P1–P3 as I have stated them: perhaps it is easy to doubt whether they are true in just their present form (particularly P1 and P2). And reflecting on the ambiguities, or plurality of uses, of

[4] E. J. Lemmon, 'Moral Dilemmas', 144–5.

various critical words or phrases ('judge better', 'want', 'intentional') it is not surprising that philosophers have tried interpreting some key phrase as meaning one thing in one principle and meaning something else in another. But I am convinced that no amount of tinkering with P1–P3 will eliminate the underlying problem: the problem will survive new wording, refinement, and elimination of ambiguity. I shall mention a few of the standard moves, and try to discredit them, but endless ways of dealing with the problem will remain. My basic strategy is therefore not that of trying to make an airtight case for P1–P3, perhaps by working them into less exceptionable form. What I hope rather is to show that P1–P3 do not contradict one another, and therefore we do not have to give up any of them. At the same time I shall offer an explanation of why we are inclined to think P1–P3 lead to a contradiction; for if I am right, a common and important mistake explains our confusion, a mistake about the nature of practical reason.

I

Here are some of the ways in which philosophers have sought, or might seek, to cope with the problem of incontinence as I have stated it.

The sins of the leopard—lust, gluttony, avarice, and wrath—are the least serious sins for which we may be eternally damned, according to Dante. Dante has these sins, which he calls the sins of incontinence, punished in the second, third, fourth, and fifth circles of Hell. In a famous example, Dante describes the adulterous sin of Francesca da Rimini and Paolo Malatesta. Commentators show their cleverness by pointing out that even in telling her story Francesca reveals her weakness of character. Thus Charles Williams says, 'Dante so manages the description, he so heightens the excuse, that the excuse reveals itself as precisely the sin . . . the persistent parleying with the occasion of sin, the sweet prolonged laziness of love . . .'[5] Perhaps all this is true of Francesca, but it is not essential to incontinence, for the 'weakness' may be momentary, not a character trait: when we speak of 'weakness' we may merely express, without explaining, the fact that the agent did what he knew to be

---

[5] C. Williams, *The Figure of Beatrice*, 118.

wrong. ('It was one page did it.') Aristotle even seems to imply that it is impossible to be habitually incontinent, on the grounds that habitual action involves a principle in accord with which one acts, while the incontinent man acts against his principle. I suppose, then, that it is at least possible to perform isolated incontinent actions, and I shall discuss incontinence as a habit or vice only as the vice is construed as the vice of often or habitually performing incontinent actions.[6]

A man might hold it to be wrong, everything considered, for him to send a valentine to Marjorie Morningstar. Yet he might send a valentine to Marjorie Eveningstar, and do it intentionally, not knowing that Marjorie Eveningstar was identical with Marjorie Morningstar. We might want to say he did something he held to be wrong, but it would be misleading to say he intentionally did something he held to be wrong; and the case I illustrate is certainly not an example of an incontinent action. We must not, I hope it is clear, think that actions can be simply sorted into the incontinent and others. 'Incontinent', like 'intentional', 'voluntary', and 'deliberate', characterizes actions only as conceived in one way rather than another. In any serious analysis of the logical form of action sentences, such words must be construed, I think, as non-truth-functional sentential operators: 'It was incontinent of Francesca that . . .' and 'It was intentional of the agent that . . .' But for present purposes it is enough to avoid the mistake of overlooking the intentionality of these expressions.

Incontinence is often characterized in one of the following ways: the agent *intends* to do *y*, which he holds to be the best course, or a better course than doing *x*; nevertheless he does *x*. Or, the agent *decides* to do *y*, which he holds to be the best course, or a better course than doing *x*, and yet he does *x*. Or, the agent *chooses y* as the result of deliberation,[7] and yet does *x*, which he deems inferior

---

[6] 'Incontinence is not strictly a vice . . . for incontinence acts against choice, vice in accord with it' (*Nic. Eth.* 1151a); 'vice is like dropsy and consumption, while incontinence is like epilepsy, vice being chronic, incontinence intermittent' (1150b). But Donne apparently describes the vice of incontinence in one of the Holy Sonnets:

> Oh, to vex me, contraries meet in one;
> Inconstancy unnaturally hath begot
> A constant habit; that when I would not
> I change in vows, and in devotion.

[7] Aristotle sometimes characterizes the incontinent man (the akrates) as 'abandoning his choice' (*Nic. Eth.*, 1151a) or 'abandoning the conclusion he has reached'

to *y*. Each of these forms of behaviour is interesting, and given some provisos may be characterized as inconsistent, weak, vacillating, or irrational. Any of them might be a case of incontinence, as I have defined it. But as they stand, they are not necessarily cases of incontinence because none of them entails that at the time he acts the agent holds that another course of action would, all things considered, be better. And on the other hand, an action can be incontinent without the agent's ever having decided, chosen, or intended to do what he judges best.

Principle 2 states a mild form of internalism. It says that a judgement of value must be reflected in wants (or desires or motives). This is not as strong as many forms of internalism: it does not, for example, say anything at all about the connection between the actual value of things (or the obligatory character of actions) and desires or motives. Nor does it, so far as I can see, involve us in any doctrine about what evaluative judgements mean. According to Hare, 'to draw attention to the close logical relations, on the one hand between wanting and thinking good, and on the other between wanting and doing something about getting what one wants, is to play into the hands of the prescriptivist; for it's to provide yet another link between thinking good and action'.[8] I confess I do not see how these 'close logical relations', which are given in one form by P1 and P2, support any particular theory about the meaning of evaluative sentences or terms. A possible source of confusion is revealed when Hare says '. . . if moral judgements were not prescriptive, there would be no problem about moral weakness; but there is a problem; therefore they are prescriptive' (p. 68). The confusion is between making a judgement, and the content of the judgement. It is P2 (or its ilk) that creates the problem, and P2 connects *making* a judgement with wanting and hence, *via* P1, with acting. But prescriptivism is a doctrine about the *content or meaning* of what is judged, and P2 says nothing about this. One could hold, for example, that to say one course of action is better than another is just to say that it will create more pleasure and yet maintain, as Mill perhaps did, that anyone who believes a certain course of action will create more pleasure than another (necessarily) wants it more. So I

(1145b); but also often along the lines suggested here: 'he does the thing he knows to be evil' (1134b) or 'he is convinced that he ought to do one thing and nevertheless does another thing' (1146b).

   [8] R. M. Hare, *Freedom and Reason*, 71.

should like to deny that there is a simple connection between the problem of incontinence as I have posed it and any particular ethical theory.

Perhaps the most common way of dealing with the problem of incontinence is to reject P2. It seems obvious enough, after all, that we may think x better, yet want y more. P2 is even easier to question if it is stated in the form: if an agent thinks he ought (or is obligated) to do x, then he wants to do x; for of course we often don't want to do what we think we ought. Hare, if I understand him, accounts for some cases of incontinence in such a way; so, according to Santas, did Plato.[9]

It is easy to interpret P2 in a way that makes it false, but it is harder to believe there is not a natural reading that makes it true. For against our tendency to agree that we often believe we ought to do something and yet don't want to, there is also the opposite tendency to say that if someone really (sincerely) believes he ought, then his belief must show itself in his behaviour (and hence, of course, in his inclination to act, or his desire). When we make a point of contrasting thinking we ought with wanting, this line continues, either we are using the phrase 'thinking we ought' to mean something like 'thinking it is what is required by the usual standards of the community' or we are restricting wanting to what is attractive on a purely selfish or personal basis. Such ways of defending P2, though I find them attractive, are hard to make conclusive without begging the present question. So I am inclined, in order to move ahead, to point out that a problem about incontinence will occur in some form as long as there is any word or phrase we can convincingly substitute for 'wants' in both P1 and P2.

Another common line to take with incontinence is to depict the akrates as overcome by passion or unstrung by emotion. 'I know indeed what evil I intend to do. But stronger than all my afterthoughts is my fury', rants Medea. Hare makes this the paradigm of all cases of weakness of the will where we cannot simply separate moral judgement and desire, and he adds that in such cases the agent is psychologically unable to do what he thinks he ought (*Freedom and Reason*, p. 77). Hare quotes Euripides' Medea when she says '. . . an unknown compulsion bears me, all reluctant, down', and St. Paul when he writes, 'The good which I want to do, I fail to

---

[9] G. Santas, 'The Socratic Paradoxes'.

do; but what I do is the wrong which is against my will; and if what I do is against my will, clearly it is no longer I who am the agent . . .' (*Romans* 7.) This line leads to the view that one never acts intentionally contrary to one's best judgement, and so denies P3; there are no incontinent actions in the sense we have defined.[10]

A related, but different, view is Aristotle's, that passion, lust, or pleasure distort judgement and so prevent an agent from forming a full-fledged judgement that his action is wrong. Though there is plenty of room for doubt as to precisely what Aristotle's view was, it is safe to say that he tried to solve our problem by distinguishing two senses in which a man may be said to know (or believe) that one thing is better than another; one sense makes P2 true, while the other sense is needed in the definition of incontinence. The flavour of this second sense is given by Aristotle's remark that the incontinent man has knowledge 'in the sense in which having knowledge does not mean knowing but only talking, as a drunken man may mutter the verses of Empedocles' (*Nic. Eth.*, 1147b).

Perhaps it is evident that there is a considerable range of actions, similar to incontinent actions in one respect or another, where we may speak of self-deception, insincerity, *mauvaise foi*, hypocrisy, unconscious desires, motives and intentions, and so on.[11] There is in fact a very great temptation, in working on this subject, to play the amateur psychologist. We are dying to say: remember the enormous variety of ways a man can believe or hold something, or know it, or want something, or be afraid of it, or do something. We can act as if we knew something, and yet profoundly doubt it; we can act at the limit of our capacity and at the same time stand off like an observer and say to ourselves, 'What an odd thing to do.' We can desire things and tell ourselves we hate them. These half-states and contradictory states are common, and full of interest to the philosopher. No doubt they explain, or at least point to a way of describing without contradiction, many cases where we find ourselves talking of weakness of the will or of incontinence. But we ourselves show a certain weak-

---

[10] Aquinas is excellent on this point. He clearly distinguishes between actions performed from a strong emotion, such as fear, which he allows are involuntary to a certain extent and hence not truly incontinent, and actions performed from concupiscence, for example: here, he says 'concupiscence inclines the will to desire the object of concupiscence. Therefore the effect of concupiscence is to make something to be voluntary.' (*Summa Theologica*, Part II, Q.6.)

[11] 'It is but a shallow haste which concludeth insincerity from what outsiders call inconsistency.' (George Eliot, *Middlemarch*.)

ness as philosophers if we do not go on to ask: does every case of incontinence involve one of the shadow-zones where we want both to apply, and to withhold, some mental predicate? Does it never happen that I have an unclouded, unwavering judgement that my action is not for the best, all things considered, and yet where the action I do perform has no hint of compulsion or of the compulsive? There is no proving such actions exist; but it seems to me absolutely certain that they do. And if this is so, no amount of attention to the subtle borderline bits of behaviour will resolve the central problem.[12]

Austin complains that in discussing the present topic, we are prone to '. . . collapse succumbing to temptation into losing control of ourselves . . .' He elaborates:

Plato, I suppose, and after him Aristotle, fastened this confusion upon us, as bad in its day and way as the later, grotesque, confusion of moral weakness with weakness of will. I am very partial to ice cream, and a bombe is served divided into segments corresponding one to one with persons at High Table: I am tempted to help myself to two segments and do, thus succumbing to temptation and even conceivably (but why necessarily?) going against my principles. But do I lose control of myself? Do I raven, do I snatch the morsels from the dish and wolf them down, impervious to the consternation of my colleagues? Not a bit of it. We often succumb to temptation with calm and even with finesse.[13]

We succumb to temptation with calm; there are also plenty of cases where we act against our better judgement and which cannot be described as succumbing to temptation.

In the usual accounts of incontinence there are, it begins to appear, two quite different themes that interweave and tend to get confused. One is, that desire distracts us from the good, or forces us to the bad; the other is that incontinent action always favours the beastly, selfish passion over the call of duty and morality. That these two themes can be separated was emphasized by Plato both in the *Protagoras* and the *Philebus* when he showed that the hedonist, on nothing but his own pleasure bent, could go against his own best

---

[12] 'Oh, tell me, who first declared, who first proclaimed, that man only does nasty things because he does not know his own real interests . . . ? What is to be done with the millions of facts that bear witness that men, knowingly, that is fully understanding their real interests, have left them in the background and have rushed headlong on another path . . . compelled to this course by nobody and by nothing . . .' (Dostoevsky, *Notes from the Underground.*)

[13] J. L. Austin, 'A Plea for Excuses', 146.

judgement as easily as anyone else. Mill makes the same point, though presumably from a position more sympathetic to the hedonist: 'Men often, from infirmity of character, make their election for the nearer good, though they know it to be the less valuable; and this no less when the choice is between two bodily pleasures than when it is between bodily and mental.' (*Utilitarianism*, Chap. 11.) Unfortunately, Mill goes on to spoil the effect of his point by adding, 'They pursue sensual indulgences to the injury of health, though perfectly aware that health is the greater good.'

As a first positive step in dealing with the problem of incontinence, I propose to divorce that problem entirely from the moralist's concern that our sense of the conventionally right may be lulled, dulled, or duped by a lively pleasure. I have just relaxed in bed after a hard day when it occurs to me that I have not brushed my teeth. Concern for my health bids me rise and brush; sensual indulgence suggests I forget my teeth for once. I weigh the alternatives in the light of the reasons: on the one hand, my teeth are strong, and at my age decay is slow. It won't matter much if I don't brush them. On the other hand, if I get up, it will spoil my calm and may result in a bad night's sleep. Everything considered I judge I would do better to stay in bed. Yet my feeling that I ought to brush my teeth is too strong for me: wearily I leave my bed and brush my teeth. My act is clearly intentional, although against my better judgement, and so is incontinent.

There are numerous occasions when immediate pleasure yields to principle, politeness, or sense of duty and yet we judge (or know) that all things considered we should opt for pleasure. In approaching the problem of incontinence it is a good idea to dwell on the cases where morality simply doesn't enter the picture as one of the contestants for our favour—or if it does, it is on the wrong side. Then we shall not succumb to the temptation to reduce incontinence to such special cases as being overcome by the beast in us, or of failing to heed the call of duty, or of succumbing to temptation.[14]

---

[14] I know no clear case of a philosopher who recognizes that incontinence is not esentially a problem in moral philosophy, but a problem in the philosophy of action. Butler, in the *Sermons* (Paragraph 39, Preface to 'The Fifteen Sermons Preached at the Rolls Chapel'), points out that 'Benevolence towards particular persons may be to a degree of weakness, and so be blamable', but here the note of self-indulgence sounds too loud. And Nowell-Smith, *Ethics*, 243ff., describes many cases of incontinence where we are overcome by conscience or duty: 'We might paradoxically, but not unfairly, say that in such a case it is difficult to resist the temptation to tell the

II

Under sceptical scrutiny, P1 and P2 appear vulnerable enough, and yet tinkering with them yields no satisfactory account of how incontinence is possible. Part of the reason at least lies in the fact that P1 and P2 derive their force from a very persuasive view of the nature of intentional action and practical reasoning. When a person acts with an intention, the following seems to be a true, if rough and incomplete, description of what goes on: he sets a positive value on some state of affairs (an end, or the performance by himself of an action satisfying certain conditions); he believes (or knows or perceives) that an action, of a kind open to him to perform, will promote or produce or realize the valued state of affairs; and so he acts (that is, he acts *because* of his value or desire and his belief). Generalized and refined, this description has seemed to many philosophers, from Aristotle on, to promise to give an analysis of what it is to act with an intention; to illuminate how we explain an action by giving the reasons the agent had in acting; and to provide the beginning of an account of practical reasoning, i.e. reasoning about what to do, reasoning that leads to action.

In the simplest case, we imagine that the agent has a desire, for example, to know the time. He realizes that by looking at his watch he will satisfy his desire; so he looks at his watch. We can answer the question why he looked at his watch; we know the intention with which he did it. Following Aristotle, the desire may be conceived as a principle of action, and its natural propositional expression would here be something like 'It would be good for me to know the time' or, even more stiffly, 'Any act of mine that results in my knowing the time is desirable.' Such a principle Aristotle compares to the major premise in a syllogism. The propositional expression of the agent's belief would in this case be, 'Looking at my watch will result in my knowing the time': this corresponds to the minor premise. Subsum-

---

truth. We are the slaves of our own consciences.' Slaves don't act freely; the case is again not clear.

Aristotle discusses the case of the man who, contrary to his own principle (and best judgement) pursues (too strongly) something noble and good (he cares too much for honour or his children), but he refuses to call this incontinence (*Nic. Eth.*, 1148).

ing the case under the rule, the agent performs the desirable action: he looks at his watch.

It seems that, given this desire and this belief, the agent is in a position to infer that looking at his watch is desirable, and in fact the making of such an inference is something it would be natural to describe as subsuming the case under the rule. But given the desire and this belief, the conditions are also satisifed that lead to (and hence explain) an intentional action, so Aristotle says that once a person has the desire and believes some action will satisfy it, *straightway he acts*. Since there is no distinguishing the conditions under which an agent is in a position to infer that an action he is free to perform is desirable from the conditions under which he acts, Aristotle apparently identifies drawing the inference and acting: he says, 'the conclusion is an action'. But of course this account of intentional action and practical reason contradicts the assumption that there are incontinent actions.

As long as we keep the general outline of Aristotle's theory before us, I think we cannot fail to realize that he can offer no satisfactory analysis of incontinent action. No doubt he can explain why, in borderline cases, we are tempted both to say an agent acted intentionally and that he knew better. But if we postulate a strong desire from which he acted, then on the theory, we also attribute to the agent a strong judgement that the action is desirable; and if we emphasize that the agent's ability to reason to the wrongness of his action was weakened or distorted, to that extent we show that he did not fully appreciate that what he was doing was undesirable.

It should not be supposed we can escape Aristotle's difficulty simply by giving up the doctrine that having the reasons for action always results in action. We might allow, for example, that a man can have a desire and believe an action will satisfy it, and yet fail to act, and add that it is only if the desire and belief cause him to act that we can speak of an intentional action.[15] On such a modified version of Aristotle's theory (if it really is a modification) we would still have to explain why in some cases the desire and belief caused an action, while in other cases they merely led to the judgement that a course of action was desirable.

The incontinent man believes it would be better on the whole to do something else, but he has a reason for what he does, for his

---

[15] For a version of this theory, see Essay 1.

action is intentional. We must therefore be able to abstract from his behaviour and state of mind a piece of practical reasoning the conclusion of which is, or would be if the conclusion were drawn from the premises, that the action actually performed is desirable. Aristotle tends to obscure this point by concentrating on cases where the incontinent man behaves 'under the influence of rule and an opinion' (*Nic. Eth.*, 1147b; cf. 1102b).

Aquinas is far clearer on this important point than Aristotle. He says:

He that has knowledge of the universal is hindered, because of a passion, from reasoning in the light of that universal, so as to draw the conclusion; but he reasons in the light of another universal proposition suggested by the inclination of the passion, and draws his conclusion accordingly . . . Hence passion fetters the reason, and hinders it from thinking and concluding under the first proposition; so that while passion lasts, the reason argues and concludes under the second.[16]

An example, given by Aquinas, shows the plight of the incontinent man:

THE SIDE OF REASON	THE SIDE OF LUST
(M_1) No fornication is lawful	(M_2) Pleasure is to be pursued
(m_1) This is an act of fornication	(m_2) This act is pleasant
(C_1) This act is not lawful	(C_2) This act is to be pursued

We can make the point more poignantly, though here we go beyond Aristotle and Aquinas, if we construe principles and conclusions as comparative judgements concerning the merits of committing, or not committing, the act in question. The two conclusions (C_1) and (C_2) will then be (given some natural assumptions): It is better not to perform this act than to perform it, and, It is better to perform this act than not to perform it. And these are in flat contradiction on the assumption that better-than is asymmetric.

And now we must observe that *this picture of moral reasoning is not merely inadequate to account for incontinence; it cannot give a correct account of simple cases of moral conflict*. By a case of moral conflict I mean a case where there are good reasons both for performing an action and for performing one that rules it out (perhaps refraining from the action). There is conflict in this mini-

[16] *Summa Theologica*, Part II, Q. 77, Art. 2, reply to objection 4. Aquinas quotes the apostle: 'I see another law in my members fighting against the law of my mind.'

mal sense whenever the agent is aware of considerations that, taken alone, would lead to mutually incompatible actions; feelings of strife and anxiety are inessential embellishments. Clearly enough, incontinence can exist only when there is conflict in this sense, for the incontinent man holds one course to be better (for a reason) and yet does something else (also for a reason). So we may set aside what is special to incontinence for the moment, and consider conflict generally. The twin arguments of the previous paragraph depict not only the plight of the incontinent man, but also of the righteous man in the toils of temptation; one of them does the wrong thing and the other the right, but both act in the face of competing claims.

The situation is common; life is crowded with examples: I ought to do it because it will save a life, I ought not because it will be a lie; if I do it, I will break my word to Lavina, if I don't, I will break my word to Lolita; and so on. Anyone may find himself in this fix, whether he be upright or temporizing, weak-willed or strong. But then unless we take the line that moral principles cannot conflict in application to a case, we must give up the concept of the nature of practical reason we have so far been assuming. For how can premises, all of which are true (or acceptable), entail a contradiction?

It is astonishing that in contemporary moral philosophy this problem has received little attention, and no satisfactory treatment. Those who recognize the difficulty seem ready to accept one of two solutions: in effect they allow only a single ultimate moral principle; or they rest happy with the notion of a distinction between the prima facie desirable (good, obligatory, etc.) and the absolutely desirable (good, obligatory, etc).[17] I shall not argue the point here, but I do not believe any version of the 'single principle' solution, once its implications are understood, can be accepted: principles, or reasons for acting, are irreducibly multiple. On the other hand, it is not easy to see how to take advantage of the purported distinction between prima facie and absolute value. Suppose first that we try to think of

---

[17] Examples of views that in effect allow only one ultimate moral principle: Kurt Baier, in *The Moral Point of View*, holds that in cases of conflict between principles there are higher-order principles that tell which principles take precedence; Singer, 'Moral Rules and Principles', claims that moral principles cannot conflict; Hare, in *The Language of Morals*, argues that there are no exceptions to acceptable moral principles. If ultimate principles never conflict or have counter-examples, we may accept the conjunction of ultimate principles as our single principle, while if there is a higher-order principle that resolves conflicts, we can obviously construct a single, exceptionless principle. And of course all outright utilitarians, rule or otherwise, believe there is a single exceptionless moral principle.

'prima facie' as an attributive adverb, helping to form such predicates as 'x is prima facie good, right, obligatory' or 'x is better, prima facie, than y'. To avoid our recent trouble, we must suppose that 'x is better, prima facie, than y' does not contradict 'y is better, prima facie, than x', and that 'x is prima facie right' does not contradict 'x is prima facie wrong'. But then the conclusion we can draw, in every case of conflict (and hence of incontinence) will be 'x is better, prima facie, than y, and y is better, prima facie, than x'. This comes down, as is clear from the structure practical reasoning would have on this assumption, to saying 'There is something to be said for, and something to be said against, doing so and so—and also for and against not doing it.' Probably this can be said about any action whatsoever; in any case it is hard to accept the idea that the sum of our moral wisdom concerning what to do in a given situation has this form. The situation I describe is not altered in any interesting way if 'prima facie' or 'prima facie obligatory' is treated as a (non-truth-functional) sentential operator rather than as a predicate. I shall return shortly to this problem; now let us reconsider incontinence.

The image we get of incontinence from Aristotle, Aquinas, and Hare is of a battle or struggle between two contestants. Each contestant is armed with his argument or principle. One side may be labelled 'passion' and the other 'reason'; they fight; one side wins, the wrong side, the side called 'passion' (or 'lust' or 'pleasure'). There is, however, a competing image (to be found in Plato, as well as in Butler and many others). It is adumbrated perhaps by Dante (who thinks he is following Aquinas and Aristotle) when he speaks of the incontinent man as one who 'lets desire pull reason from her throne' (*Inferno*, Canto v). Here there are three actors on the stage: reason, desire, and the one who lets desire get the upper hand. The third actor is perhaps named 'The Will' (or 'Conscience'). It is up to The Will to decide who wins the battle. If The Will is strong, he gives the palm to reason; if he is weak, he may allow pleasure or passion the upper hand.

This second image is, I suggest, superior to the first, absurd as we may find both. On the first story, not only can we not account for incontinence; it is not clear how we can ever blame the agent for what he does: his action merely reflects the outcome of a struggle within him. What could he do about it? And more important, the first image does not allow us to make sense of a conflict in one person's soul, for it leaves no room for the all-important process of weighing considerations.[18] In the second image, the agent's representative, The Will, can judge the strength of the arguments on both sides, can execute the decision, and take the rap. The only trouble is that we seem back where we started. For how can The Will judge one course of action better and yet choose the other?

It would be a mistake to think we have made no progress. For what these colourful gladiatorial and judicial metaphors ought now to suggest is that there is a piece of practical reasoning present in moral conflict, and hence in incontinence, which we have so far entirely neglected. What must be added to the picture is a new argument:

THE WILL (CONSCIENCE)

(M_3) M_1 and M_2

(m_3) m_1 and m_2

(C_3) This action is wrong

Clearly something like this third argument is necessary if an agent is to act either rightly or incontinently in the face of conflict. It is not enough to know the reasons on each side: he must know how they add up.[19] The incontinent man goes against his better judgement, and this surely is (C_3), which is based on all the considerations, and not (C_1) which fails to bring in the reasons on the other side. You could say we have discovered two quite different meanings of the phrase 'his better judgement'. It might mean, any judgement for the right side (reason, morality, family, country); or, the judgement based on all relevant considerations known to the actor. The first notion, I have argued, is really irrelevant to the analysis of incontinence.

But now we are brought up against our other problem, the form or nature of practical reasoning. For nothing could be more obvious than that our third 'practical syllogism' is no syllogism at all; the conclusion simply doesn't follow by logic from the premises. And introducing the third piece of reasoning doesn't solve the problem we had before anyway: we still have contradictory 'conclusions'. We

[18] A more sophisticated account of conflict that seems to raise the same problem is Ryle's account in *The Concept of Mind*, 93–5.

[19] My authority for how they do add up in this case is Aquinas: see the reference in footnote 16.

could at this point try once more introducing 'prima facie' in suitable places: for example, in (M_1), (M_2), (C_1), and (C_2). We might then try to relate prima facie desirability to desirability sans phrase by making (C_1) and (C_2), thus interpreted, the data for (C_3). But this is an unpromising line. We can hardly expect to learn whether an action ought to be performed simply from the fact that it is both prima facie right and prima facie wrong.

The real source of difficulty is now apparent: if we are to have a coherent theory of practical reason, we must give up the idea that we can *detach* conclusions about what is desirable (or better) or obligatory from the principles that lend those conclusions colour. The trouble lies in the tacit assumption that moral principles have the form of universalized conditionals; once this assumption is made, nothing we can do with a prima facie operator in the conclusion will save things. The situation is, in this respect, like reasoning from probabilistic evidence. As Hempel has emphasized with great clarity,[20] we cannot reason from:

(M_4) If the barometer falls, it almost certainly will rain
(m_4) The barometer is falling

to the conclusion:

(C_4) It almost certainly will rain

since we may at the same time be equally justified in arguing:

(M_5) Red skies at night, it almost certainly won't rain
(m_5) The sky is red tonight
∴ (C_5) It almost certainly won't rain

The crucial blunder is interpreting (M_4) and (M_5) to allow detachment of the modal conclusion. A way to mend matters is to view the 'almost certainly' of (M_4) and (M_5) as modifying, not the conclusion, but the connective. Thus we might render (M_4), 'That the barometer falls probabilizes that it will rain'; in symbols, '$pr(Rx, Fx)$', where the variable ranges over areas of space-time that may be characterized by falling barometers or rain. If we let 'a' name the space and time of here and now, and 'Sx' mean that the early part of x is characterized by a red sky of evening, we may attempt to reconstruct the thought bungled above thus:

[20] Carl Hempel, *Aspects of Scientific Explanation*, 394–403.

$pr(Rx,Fx)$ $pr(\sim Rx,Sx)$
Fa Sa
$\therefore\ pr(Ra,pr(Rx,Fx))$ and Fa $\therefore\ pr(\sim Ra,pr(\sim Rx,Sx))$ and Sa

If we want to predict the weather, we will take a special interest in:

$pr(\sim Ra,e)$ or $pr(Ra,e)$

where e is all the relevant evidence we have. But it is clear that we can infer neither of these from the two arguments that went before, even if e is simply the conjunction of their premises (and even if for our qualitative 'pr' we substitute a numerical measure of degree of support).

I propose to apply the pattern to practical reasoning in the obvious way. The central idea is that a moral principle, like 'Lying is (prima facie) wrong', cannot coherently be treated as a universally quantified conditional, but should be recognized to mean something like, 'That an act is a lie prima facie makes it wrong'; in symbols, '$pf(Wx,Lx)$'. The concept of the prima facie, as it is needed in moral philosophy, relates propositions. In logical grammar, 'prima facie' is not an operator on single sentences, much less on predicates of actions, but on pairs of sentences related as (expressing) moral judgement and ground. Here is how the piece of practical reasoning misrepresented by (M_1), (m_1) and (C_1) might look when reconstituted:

(M_6) $pf(x$ is better than y, x is a refraining from fornication and y is an act of fornication)

(m_6) a is a refraining from fornication and b is an act of fornication

$\therefore\ (C_6)$ $pf(a$ is better than b, (M_6) and $(m_6))$

Similarly, (M_2) and (m_2), when rewritten in the new mode, and labelled (M_7) and (m_7), will yield:

(C_7) $pf(b$ is better than a, (M_7) and $(m_7))$

A judgement in which we will take particular interest is:

(C_8) $pf(a$ is better than b, $e)$

where e is all the relevant considerations known to us, including at least (M_6), (m_6), (M_7), and (m_7).

Of course (C_8) does not follow logically from anything that went before, but in this respect moral reasoning seems no worse off than predicting the weather. In neither case do we know a general formula for computing how far or whether a conjunction of evidence statements supports a conclusion from how far or whether each conjunct supports it. There is no loss either, in this respect, in our strategy of relativizing moral judgements: we have no clue how to arrive at (C_8) from the reasons, but its faulty prototype (C_3) was in no better shape. There has, however, been a loss of relevance, for the conditionalization that keeps (C_6) from clashing with (C_7), and (C_8) from clashing with either, also insulates all three from action. Intentional action, I have argued in defending P1 and P2, is geared directly to unconditional judgements like 'It would be better to do *a* than to do *b*.' Reasoning that stops at conditional judgements such as (C_8) is practical only in its subject, not in its issue.[21]

Practical reasoning does however often arrive at unconditional judgements that one action is better than another—otherwise there would be no such thing as acting on a reason. The minimal elements of such reasoning are these: the agent accepts some reason (or set of reasons) *r*, and holds that *pf(a* is better than *b, r)*, and these constitute the reason why he judges that *a* is better than *b*. Under these conditions, the agent will do *a* if he does either *a* or *b* intentionally, and his reason for doing *a* rather than *b* will be identical with the reason why he judges *a* better than *b*.

This modified account of acting on a reason leaves P1 and P2 untouched, and Aristotle's remark that the conclusion (of a piece of practical reasoning) is an action remains cogent. But now there is no (logical) difficulty in the fact of incontinence, for the akrates is characterized as holding that, all things considered, it would be better to do *b* than to do *a*, even though he does *a* rather than *b* and with a reason. The logical difficulty has vanished because a judgement that *a* is better than *b*, all things considered, is a relational, or *pf*, judgement, and so cannot conflict logically with any unconditional judgement.

Possibly it will be granted that P1–P3, as interpreted here, do not yield a contradiction. But at the same time, a doubt may arise whether P3 is plausible, given this interpretation. For how is it possible for a man to judge that *a* is better than *b*, all things considered, and not judge that *a* is better than *b*?

---

[21] This claim is further pursued in Essay 5.

One potential confusion is quickly set aside. '*a* is better than *b*, all things (viz. all truths, moral and otherwise) considered' surely does entail '*a* is better than *b*', and we do not want to explain incontinence as a simple logical blunder. The phrase 'all things considered' must, of course, refer only to things known, believed, or held by the agent, the sum of his relevant principles, opinions, attitudes, and desires. Setting this straight may, however, seem only to emphasize the real difficulty. We want now to ask: how is it possible for a man to judge that *a* is better than *b* on the grounds that *r*, and yet not judge that *a* is better than *b*, when *r* is the sum of all that seems relevant to him? When we say that *r* contains all that seems relevant to the agent, don't we just mean that nothing has been omitted that influences his judgement that *a* is better than *b*?

Since what is central to the solution of the problem of incontinence proposed in this paper is the contrast between conditional (prima facie) evaluative judgements and evaluative judgements sans phrase, perhaps we can give a characterization of incontinence that avoids the troublesome 'all things considered'. A plausible modification of our original defintion (D) of incontinence might label an action, *x*, as incontinent provided simply that the agent has a better reason for doing something else: he does *x* for a reason *r*, but he has a reason *r'* that includes *r* and more, on the basis of which he judges some alternative *y* to be better than *x*.[22] Of course it might also have been incontinent of him to have done *y*, since he may have had a better reason still for performing some third action *z*. Following this line, we might say that an action *x* is continent if *x* is done for a reason *r*, and there is no reason *r'* (that includes *r*), on the basis of which the agent judges some action better than *x*.

This shows we can make sense of incontinence without appeal to the idea of an agent's total wisdom, and the new formulation might in any case be considered an improvement on (D) since it allows (correctly, I think) that there are incontinent actions even when no judgement is made in the light of all the reasons. Still, we cannot rule out the case where a judgement is made in the light of all the reasons, so the underlying difficulty may be thought to remain.

In fact, however, the difficulty is not real. Every judgement is made in the light of all the reasons in this sense, that it is made in the presence of, and is conditioned by, that totality. But this does not

---

[22] We might want to rule out the case, allowed by this formulation, where the agent does what he has the best reason for doing, but does not do it for that reason.

mean that every judgement is reasonable, or thought to be so by the agent, on the basis of those reasons, nor that the judgement was reached from that basis by a process of reasoning. There is no paradox in supposing a person sometimes holds that all that he believes and values supports a certain course of action, when at the same time those same beliefs and values cause him to reject that course of action.[23] If r is someone's reason for holding that p, then his holding that r must be, I think, a cause of his holding that p. But, and this is what is crucial here, his holding that r may cause his holding that p without r being his reason; indeed, the agent may even think that r is a reason to reject p.

It is possible, then, to be incontinent, even if P1 and P2 are true. But what, on this analysis, is the fault in incontinence? The akrates does not, as is now clear, hold logically contradictory beliefs, nor is his failure necessarily a moral failure. What is wrong is that the incontinent man acts, and judges, irrationally, for this is surely what we must say of a man who goes against his own best judgement. Carnap and Hempel have argued that there is a principle which is no part of the logic of inductive (or statistical) reasoning, but is a directive the rational man will accept. It is the *requirement of total evidence for inductive reasoning*: give your credence to the hypothesis supported by all available relevant evidence.[24] There is, I suggest, an analogous principle the rational man will accept in applying practical reasoning: perform the action judged best on the basis of all available relevant reasons. It would be appropriate to call this the *principle of continence*. There may seem something queer in making the requirement of total evidence an imperative (can one tailor one's beliefs to order?), but there is no such awkwardness about the principle of continence. It exhorts us to actions we can perform if we want; it leaves the motives to us. What is hard is to acquire the virtue of continence, to make the principle of continence our own. But there is no reason in principle why it is any more difficult to become continent than to become chaste or brave. One gets a lively sense of the difficulties in St. Augustine's extraordinary prayer: 'Give me chastity and continence, only not yet' (*Confessions*, VIII, vii).

[23] At this point my account of incontinence seems to me very close to Aristotle's. See G. E. M. Anscombe, 'Thought and Action in Aristotle'.

[24] See Hempel, op. cit., 397–403 for important modifications, and further references.

Why would anyone ever perform an action when he thought that, everything considered, another action would be better? If this is a request for a psychological explanation, then the answers will no doubt refer to the interesting phenomena familiar from most discussions of incontinence: self-deception, overpowering desires, lack of imagination, and the rest. But if the question is read, what is the agent's reason for doing *a* when he believes it would be better, all things considered, to do another thing, then the answer must be: for this, the agent has no reason.[25] We perceive a creature as rational in so far as we are able to view his movements as part of a rational pattern comprising also thoughts, desires, emotions, and volitions. (In this we are much aided by the actions we conceive to be utterances.) Through faulty inference, incomplete evidence, lack of diligence, or flagging sympathy, we often enough fail to detect a pattern that is there. But in the case of incontinence, the attempt to read reason into behaviour is necessarily subject to a degree of frustration.

What is special in incontinence is that the actor cannot understand himself: he recognizes, in his own intentional behaviour, something essentially surd.

[25] Of course he has a reason for doing *a*; what he lacks is a reason for not letting his better reason for not doing *a* prevail.

3 Agency

What events in the life of a person reveal agency; what are his deeds and his doings in contrast to mere happenings in his history; what is the mark that distinguishes his actions?

This morning I was awakened by the sound of someone practising the violin. I dozed a bit, then got up, washed, shaved, dressed, and went downstairs, turning off a light in the hall as I passed. I poured myself some coffee, stumbling on the edge of the dining room rug, and spilled my coffee fumbling for the *New York Times*.

Some of these items record things I did; others, things that befell me, things that happened to me on the way to the dining room. Among the things I did were get up, wash, shave, go downstairs, and spill my coffee. Among the things that happened to me were being awakened and stumbling on the edge of the rug. A borderline case, perhaps, is dozing. Doubts could be kindled about other cases by embroidering on the story. Stumbling can be deliberate, and when so counts as a thing done. I might have turned off the light by inadvertently brushing against the switch; would it then have been my deed, or even something that I did?

Many examples can be settled out of hand, and this encourages the hope that there is an interesting principle at work, a principle which, if made explicit, might help explain why the difficult cases are difficult. On the other side a host of cases raise difficulties. The question itself seems to go out of focus when we start putting pressure on such phrases as 'what he did', 'his actions', 'what happened to him', and it often matters to the appropriateness of the answer what form we give the question. (Waking up is something I did, perhaps, but not an action.) We should maintain a lively sense of the possibility that the question with which we began is, as Austin suggested, a misguided one.[1]

In this essay, however, I once more try the positive assumption, that the question is a good one, that there is a fairly definite subclass of events which are actions. The costs of the assumption are the usual ones: oversimplification, the setting aside of large classes of exceptions, the neglect of distinctions hinted by grammar and common sense, recourse to disguised linguistic legislation. With luck we learn something from such methods. There may, after all, be important and general truths in this area, and if there are how else will we discover them?

Philosophers often seem to think that there must be some simple grammatical litmus of agency, but none has been discovered. I drugged the sentry, I contracted malaria, I danced, I swooned, Jones was kicked by me, Smith was outlived by me: this is a series of examples designed to show that a person named as subject in sentences in the active (whether or not the verb is transitive) or as object in sentences in the passive, may or may not be the agent of the event recorded.[2]

Another common error is to think verbs may be listed according to whether they do or do not impute agency to a subject or object. What invites the error is that this is true of some verbs. To say of a person that he blundered, insulted his uncle, or sank the *Bismarck* is automatically to convict him of being the author of those events; and to mention someone in the subject position in a sentence with the verb in the passive tense is, so far as I can see, to ensure that he is not the agent. But very often a sentence will record an episode in the life of the agent and leave us in the dark as to whether it was an action. Here are some examples: he blinked, rolled out of bed, turned on the light, coughed, squinted, sweated, spilled the coffee, and tripped over the rug. We know whether these events are actions only after we know more than the verb tells us. By considering the additional information that would settle the matter, we may find an answer to the question of what makes a bit of biography an action.

One hint was given in my opening fragmentary diary. Tripping over a rug is normally not an action; but it is if done intentionally. Perhaps, then, being intentional is the relevant distinguishing mark.

---

[1] See J. L. Austin, 'A Plea for Excuses', 126–7.
[2] The point is developed in Irving Thalberg's 'Verbs, Deeds and What Happens to Us', 259–60.

If it were, it would help explain why some verbs imply agency, for some verbs describe actions that cannot be anything but intentional; asserting, cheating, taking a square root, and lying are examples.

This mark is not sufficient, however, for although intention implies agency, the converse does not hold. Thus spilling the coffee, sinking the *Bismarck*, and insulting someone are all things that may or may not be done intentionally, but even when not intentional, they are normally actions. If, for example, I intentionally spill the contents of my cup, mistakenly thinking it is tea when it is coffee, then spilling the coffee is something I do, it is an action of mine, though I do not do it intentionally. On the other hand, if I spill the coffee because you jiggle my hand, I cannot be called the agent. Yet while I may hasten to add my excuse, it is not incorrect, even in this case, to say I spilled the coffee. Thus we must distinguish three situations in which it is correct to say I spilled the coffee: in the first, I do it intentionally; in the second I do not do it intentionally but it is my action (I thought it was tea); in the third it is not my action at all (you jiggle my hand).[3]

Certain kinds of mistake are particularly interesting: misreading a sign, misinterpreting an order, underestimating a weight, or miscalculating a sum. These are things that strictly speaking cannot be done intentionally. One can pretend to misread a sign, one can underestimate a weight through sloth or inattention, or deliberately write down what one knows to be a wrong answer to an addition; but none of these is an intentional flubbing. To make a mistake of one of the mentioned kinds is to fail to do what one intends, and one cannot, Freudian paradox aside, intend to fail. These mistakes are not intentional, then; nevertheless, they are actions. To see this we need only notice that making a mistake must in each case be doing something else intentionally. A misreading must be a reading, albeit one that falls short of what was wanted; misinterpreting an order is a case of interpreting it (and with the intention of getting it right); underestimating is estimating; and a miscalculation is a calculation (though one that founders).

Can we now say which events involve agency? Intentional actions do, and so do some other things we do. What is the common element? Consider coffee spilling again. I am the agent if I spill the

---

[3] This threefold division should not be confused with Austin's subtle work on the differences among purpose, intention, and deliberation in 'Three Ways of Spilling Ink'.

coffee meaning to spill the tea, but not if you jiggle my hand. What is the difference? The difference seems to lie in the fact that in one case, but not in the other, I am intentionally doing *something*. My spilling the contents of my cup was intentional; as it happens, this very same act can be redescribed as my spilling the coffee. Of course, thus redescribed the action is no longer intentional; but this fact is apparently irrelevant to the question of agency.

And so I think we have one correct answer to our problem: a man is the agent of an act if what he does can be described under an aspect that makes it intentional.

What makes this answer possible is the semantic opacity, or intentionality, of attributions of intention. Hamlet intentionally kills the man behind the arras, but he does not intentionally kill Polonius. Yet Polonius is the man behind the arras, and so Hamlet's killing of the man behind the arras is identical with his killing of Polonius. It is a mistake to suppose there is a class of intentional actions: if we took this tack, we should be compelled to say that one and the same action was both intentional and not intentional. As a first step toward straightening things out, we may try talking not of actions but of sentences and descriptions of actions instead. In the case of agency, my proposal might then be put: a person is the agent of an event if and only if there is a description of what he did that makes true a sentence that says he did it intentionally. This formulation, with its quantification over linguistic entities, cannot be considered entirely satisfactory. But to do better would require a semantic analysis of sentences about propositional attitudes.[4]

Setting aside the need for further refinement, the proposed criterion of actions seems to fit the examples we have discussed. Suppose an officer aims a torpedo at a ship he thinks is the *Tirpitz* and actually sinks the *Bismarck*. Then sinking the *Bismarck* is his action, for that action is identical with his attempt to sink the ship he took to be the *Tirpitz*, which is intentional. Similarly, spilling the coffee is the act of a person who does it by intentionally spilling the contents of his cup. Also it is clearer now why mistakes are actions, for making a mistake must be doing something with the intention of achieving a result that is not forthcoming.

If we can say, as I am urging, that a person does, as agent, whatever he does intentionally under some description, then,

---

[4] For an attempt at such a theory see my 'On Saying That', and further discussion in Essay 6.

although the *criterion* of agency is, in the semantic sense, intentional, the *expression* of agency is itself purely extensional. The relation that holds between a person and an event, when the event is an action performed by the person, holds regardless of how the terms are described. Therefore we can without confusion speak of the class of events that are actions, which we cannot do with intentional actions.

It might be thought that the concept of an action is hopelessly indistinct because it is so hard to decide whether or not knocking over a policeman, say, or falling downstairs, or deflating someone's ego is an action. But if being an action is a trait which particular events have independently of how they are described, there is no reason to expect in general to be able to tell, merely by knowing some trait of an event (that it is a case of knocking over a policeman, say), whether or not it is an action.

Is our criterion so broad that it will include under actions events that no one would normally count as actions? For example, isn't tripping over the edge of the rug just part of my intentional progress into the dining room? I think not. An intentional movement of mine did cause me to trip, and so I did trip myself: this was an action, though not an intentional one. But 'I tripped' and 'I tripped myself' do not report the same event. The first sentence is entailed by the second, because to trip myself is to do something that results in my tripping; but of course doing something that results in my tripping is not identical with what it causes.

The extensionality of the expression of agency suggests that the concept of agency is simpler or more basic than that of intention, but unfortunately the route we have travelled does not show how to exploit the hint, for all we have so far is a way of picking out cases of agency by appeal to the notion of intention. This is to analyse the obscure by appeal to the more obscure—not as pointless a process as it is often thought to be, but still disappointing. We should try to see if we can find a mark of agency that does not use the concept of intention.

The notion of cause may provide the clue. With respect to causation, there is a certain rough symmetry between intention and agency. If I say that Smith set the house on fire in order to collect the insurance, I explain his action, in part, by giving one of its causes, namely his desire to collect the insurance. If I say that Smith burned down the house by setting fire to the bedding, then I explain the conflagration by giving a cause, namely Smith's action. In both cases, causal explanation takes the form of a fuller description of an action, either in terms of a cause or of an effect. To describe an action as one that had a certain purpose or intended outcome is to describe it as an effect; to describe it as an action that had a certain outcome is to describe it as a cause. Attributions of intention are typically excuses and justifications; attributions of agency are typically accusations or assignments of responsibility. Of course the two kinds of attribution do not rule one another out, since to give the intention with which an act was done is also, and necessarily, to attribute agency. If Brutus murdered Caesar with the intention of removing a tyrant, then a cause of his action was a desire to remove a tyrant and an effect was the death of Caesar. If the officer sank the *Bismarck* with the intention of sinking the *Tirpitz*, then an action of his was caused by his desire to sink the *Tirpitz* and had the consequence that the *Bismarck* sank.[5]

These examples and others suggest that, in every instance of action, the agent made happen or brought about or produced or authored the event of which he was the agent, and these phrases in turn seem grounded in the idea of cause. Can we then say that to be the author or agent of an event is to cause it? This view, or something apparently much like it, has been proposed or assumed by a number of recent authors.[6] So we should consider whether introducing the notion of causation in this way can improve our understanding of the concept of agency.

Clearly it can, at least up to a point. For an important way of justifying an attribution of agency is by showing that some event was caused by something the agent did. If I poison someone's morning grapefruit with the intention of killing him, and I succeed, then I caused his death by putting poison in his food, and that is why I am the agent in his murder. When I manage to hurt someone's feelings by denigrating his necktie, I cause the hurt, but it is another event, my saying something mean, that is the cause of the hurt.

---

[5] In Essay 1, I develop the theme that to give a reason or intention with which an action is performed is, among other things, to describe the action in terms of a cause. In this essay I explore how the effects of actions enter into our descriptions of them.

[6] For example, Roderick Chisholm, 'Freedom and Action', Daniel Bennett, 'Action, Reason and Purpose', Anthony Kenny, *Action, Emotion and Will*, Georg Henrick von Wright, *Norm and Action*, Richard Taylor, *Action and Purpose*. Further criticism of this kind of causal analysis of agency can be found in Irving Thalberg, 'Do We Cause our Own Actions?' and in Essay 6.

The notion of cause appealed to here is ordinary event causality, the relation, whatever it is, that holds between two events when one is cause of the other. For although we say the agent caused the death of the victim, that is, that he killed him, this is an elliptical way of saying that some act of the agent—something he did, such as put poison in the grapefruit—caused the death of the victim.

Not every event we attribute to an agent can be explained as caused by another event of which he is agent: some acts must be primitive in the sense that they cannot be analysed in terms of their causal relations to acts of the same agent. But then event causality cannot in this way be used to explain the relation between an agent and a primitive action. Event causality can spread responsibility for an action to the consequences of the action, but it cannot help explicate the first attribution of agency on which the rest depend.[7]

If we interpret the idea of a bodily movement generously, a case can be made for saying that all primitive actions are bodily movements. The generosity must be openhanded enough to encompass such 'movements' as standing fast, and mental acts like deciding and computing. I do not plan to discuss these difficult examples now; if I am wrong about the precise scope of primitive actions, it will not affect my main argument. It is important, however, to show that in such ordinary actions as pointing one's finger or tying one's shoelaces the primitive action is a bodily movement.

I can imagine at least two objections to this claim. First, it may be said that in order to point my finger, I do something that causes the finger to move, namely contract certain muscles; and perhaps this requires that I make certain events take place in my brain. But these events do not sound like ordinary bodily movements. I think that the premises of this argument may be true, but that the conclusion does not follow. It may be true that I cause my finger to move by contracting certain muscles, and possibly I cause the muscles to contract by making an event occur in my brain. But this does not show that pointing my finger is not a primitive action, for it does not show that I must do something else that causes it. Doing something

---

[7] Here, and in what follows, I assume that we have set aside an analysis of agency that begins by analysing the concept of intention, or of acting with an intention, or of a reason in acting. These concepts can be analysed, at least in part, in terms of event causality. In Essay 1, I try to show that although beliefs and desires (and similar mental states) are not events, we can properly say that they are causes of intentional actions, and when we say this we draw upon the concept of ordinary event causality.

that causes my finger to move does not cause me to move my finger; it *is* moving my finger.

In discussing examples like this one, Chisholm has suggested that, although an agent may be said to make certain cerebral events happen when it is these events that cause his finger to move, making the cerebral events happen cannot be called something that he does. Chisholm also thinks that many things an agent causes to happen, in the sense that they are events caused by things he does, are not events of which he is the agent. Thus if moving his finger is some-thing a man does, and this movement causes some molecules of air to move, then although the man may be said to have caused the molecules to move, and hence to have moved the molecules, this is not something he did.[8]

It does not seem to me that this is a clear or useful distinction: all of Chisholm's cases of making something happen are, so far as my intuition goes, cases of agency, situations in which we may, and do, allow that the person did something, that he was an agent. When a person makes an event occur in his brain, he does not normally know that he is doing this, and Chisholm seems to suggest that for this reason we cannot say it is something that he does. But a man may even be doing something intentionally and not know that he is; so of course he can be doing it without knowing that he is. (A man may be making ten carbon copies as he writes, and this may be intentional; yet he may not know that he is; all he knows is that he is trying.)

Action does require that what the agent does is intentional under some description, and this in turn requires, I think, that what the agent does is known to him under some description. But this condi-tion is met by our examples. A man who raises his arm both intends to do with his body whatever is needed to make his arm go up and knows that he is doing so. And of course the cerebral events and movements of the muscles are just what is needed. So, though the agent may not know the names or locations of the relevant muscles, nor even know he has a brain, what he makes happen in his brain and muscles when he moves his arm is, under one natural descrip-tion, something he intends and knows about.

The second objection to the claim that primitive actions are bodily movements comes from the opposite direction: it is that some primitive actions involve more than a movement of the body. When I tie my shoelaces, there is on the one hand the movement of my fingers, and on the other the movement of the laces. But is it possible to separate these events by calling the first, alone, my action? What makes the separation a problem is that I do not seem able to describe or think how I move my fingers, apart from moving the laces. I do not move my fingers in the attempt to cause my shoes to be tied, nor am I capable of moving my fingers in the appropriate way when no laces are present (this is a trick I might learn). Similarly, it might be argued that when they utter words most people do not know what muscles to move or how to hold their mouths in order to produce the words they want; so here again it seems that a primitive action must include more than a bodily movement, namely a motion of the air.

The objection founders for the same reason as the last one. Everything depends on whether or not there is an appropriate description of the action. It is correctly assumed that unless the agent himself is aware of what he is doing with his body alone, unless he can conceive his movements as an event physically separate from whatever else takes place, his bodily movements cannot be his action. But it is wrongly supposed that such awareness and conception are impossible in the case of speaking or of tying one's shoelaces. For an agent always knows how he moves his body when, in acting intentionally, he moves his body, in the sense that there is *some* description of the movement under which he knows that he makes it. Such descriptions are, to be sure, apt to be trivial and unrevealing; this is what ensures their existence. So, if I tie my shoelaces, here is a description of my movements: I move my body in just the way required to tie my shoelaces. Similarly, when I utter words, it is true that I am unable to describe what my tongue and mouth do, or to name the muscles I move. But I do not need the terminology of the speech therapist: what I do is move my mouth and muscles, as I know how to do, in just the way needed to produce the words I have in mind.

So there is, after all, no trouble in producing familiar and correct descriptions of my bodily movements, and these are the events that cause such further events as my shoelaces' being tied or the air's vibrating with my words. Of course, the describing trick has been turned by describing the actions as the movements with the right effects; but this does not show the trick has not been turned. What was needed was not a description that did not mention the effects, but a description that fitted the cause. There is, I conclude, nothing standing in the way of saying that our primitive actions, at least if we set aside such troublesome cases as mental acts, are bodily movements.

To return to the question whether the concept of action may be analysed in terms of the concept of causality: what our discussion has shown is that we may concentrate on primitive actions. The ordinary notion of event causality is useful in explaining how agency can spread from primitive actions to actions described in further ways, but it cannot in the same way explain the basic sense of agency. What we must ask, then, is whether there is another kind of causality, one that does not reduce to event causality, an appeal to which will help us understand agency. We may call this kind of causality (following Thalberg) *agent causality*.

If we restrict ourselves, for the reason just given, to primitive actions, how well does the idea of agent causality account for the relation between an agent and his action? There is this dilemma: either the causing by an agent of a primitive action is an event discrete from the primitive action, in which case we have problems about acts of the will or worse, or it is not a discrete event, in which case there seems no difference between saying someone caused a primitive action and saying he was the agent.

To take the first horn: suppose that causing a primitive action (in the sense of agent causality) does introduce an event separate from, and presumably prior to, the action. This prior event in turn must either be an action, or not. If an action, then the action we began with was not, contrary to our assumption, primitive. If not an action, then we have tried to explain agency by appeal to an even more obscure notion, that of a causing that is not a doing.

One is impaled on the second horn of the dilemma if one supposes that agent causation does *not* introduce an event in addition to the primitive action. For then what more have we said when we say the agent caused the action than when we say he was the agent of the action? The concept of *cause* seems to play no role. We may fail to detect the vacuity of this suggestion because causality does, as we have noticed, enter conspicuously into accounts of agency; but where it does it is the garden-variety of causality, which sheds no light on the relation between the agent and his primitive actions.

We explain a broken window by saying that a brick broke it; what explanatory power the remark has derives from the fact that we may first expand the account of the cause to embrace an event, the movement of the brick, and we can then summon up evidence for the existence of a law connecting such events as motions of medium-sized rigid objects and the breaking of windows. The ordinary notion of cause is inseparable from this elementary form of explanation. But the concept of agent causation lacks these features entirely. What distinguishes agent causation from ordinary causation is that no expansion into a tale of two events is possible, and no law lurks. By the same token, nothing is explained. There seems no good reason, therefore, for using such expressions as 'cause', 'bring about', 'make the case' to *illuminate* the relation between an agent and his act. I do not mean that there is anything wrong with such expressions—there are times when they come naturally in talk of agency. But I do not think that by introducing them we make any progress towards understanding agency and action.

Causality is central to the concept of agency, but it is ordinary causality between events that is relevant, and it concerns the effects and not the causes of actions (discounting, as before, the possibility of analysing intention in terms of causality). One way to bring this out is by describing what Joel Feinberg calls the 'accordion effect',[9] which is an important feature of the language we use to describe actions. A man moves his finger, let us say intentionally, thus flicking the switch, causing a light to come on, the room to be illuminated, and a prowler to be alerted. This statement has the following entailments: the man flicked the switch, turned on the light, illuminated the room, and alerted the prowler. Some of these things he did intentionally, some not; beyond the finger movement, intention is irrelevant to the inferences, and even there it is required only in the sense that the movement must be intentional under some description. In brief, once he has done one thing (move a finger), each consequence presents us with a deed; an agent causes what his actions cause.[10]

[9] Joel Feinberg, 'Action and Responsibility'.
[10] The formulation in this sentence is more accurate than in some of my examples. Suppose Jones intentionally causes Smith intentionally to shoot Clifford to death. We certainly won't conclude that Jones shot Clifford, and we may or may not say that Jones killed Clifford. Still, my formulation is correct provided we can go from 'Jones's action caused Clifford's death' to 'Jones caused Clifford's death'. There will, of course, be a conflict if we deny that both Jones and Smith (in our story) could be said to have caused Clifford's death, and at the same time affirm the transitivity of

The accordion effect will not reveal in what respect an act is intentional. If someone moves his mouth in such a way as to produce the words 'your bat is on hackwards', thus causing offence to his companion, the accordion effect applies, for we may say both that he spoke those words and that he offended his companion. Yet it is possible that he did not intend to move his mouth so as to produce those words, nor to produce them, nor to offend his companion. But the accordion effect is not applicable if there is no intention present. If the officer presses a button thinking it will ring a bell that summons a steward to bring him a cup of tea, but in fact it fires a torpedo that sinks the *Bismarck*, then the officer sank the *Bismarck*; but if he fell against the button because a wave upset his balance, then, though the consequences are the same, we will not count him as the agent.

The accordion effect is limited to agents. If Jones intentionally swings a bat that strikes a ball that hits and breaks a window, then Jones not only struck the ball but also broke the window. But we do not say that the bat, or even its movement, broke the window, though of course the movement of the bat caused the breakage. We do indeed allow that inanimate objects cause or bring about various things—in our example, the ball did break the window. However, this is not the accordion effect of agency, but only the ellipsis of event causality. The ball broke the window—that is to say, its motion caused the breakage.

It seems therefore that we may take the accordion effect as a mark of agency. It is a way of inquiring whether an event is a case of agency to ask whether we can attribute its effects to a person. And on the other hand, whenever we say a person has done something where what we mention is clearly not a bodily movement, we have made him the agent not only of the mentioned event, but of some bodily movement that brought it about. In the case of bodily movements we sometimes have a brief way of mentioning a person and an event and yet of leaving open the question of whether he was the agent, as: Smith fell down.

The accordion effect is interesting because it shows that we treat the consequences of actions differently from the way we treat the

causality. We could, however, preserve the formula in the face of a denial that under the circumstances Jones could be said to have caused Clifford's death by saying that under the circumstances the transitivity of causality also breaks down. For further discussion of the issue, see H. L. A. Hart and A. M. Honoré, *Causation in the Law*, Joel Feinberg, 'Causing Voluntary Actions', and J. E. Atwell, 'The Accordion-Effect Thesis'.

consequences of other events. This suggests that there is, after all, a fairly simple linguistic test that sometimes reveals that we take an event to be an action. But as a criterion it can hardly be counted as satisfactory: it works for some cases only, and of course it gives no clue as to what makes a primitive action an action.

At this point I abandon the search for an analysis of the concept of agency that does not appeal to intention, and turn to a related question that has come to the fore in the discussion of agent causality and the accordion effect. The new question is what relation an agent has to those of his actions that are not primitive, those actions in describing which we go beyond mere movements of the body and dwell on the consequences, on what the agent has wrought in the world beyond his skin. Assuming that we understand agency in the case of primitive actions, how exactly are such actions related to the rest? The question I now raise may seem already to have been settled, but in fact it has not. What *is* clear is the relation between a primitive action, say moving one's finger in a certain way, and a consequence such as one's shoelaces being tied: it is the relation of event causality. But this does not give a clear answer to the question of how the movement of the hands is related to the action of tying one's shoelaces, nor for that matter, to the question of how the action of tying one's shoelaces is related to one's shoelaces being tied. Or, to alter the example, if Brutus killed Caesar by stabbing him, what is the relation between these two actions, the relation expressed by the 'by'? No doubt it is true that Brutus killed Caesar because the stabbing resulted in Caesar's death; but we still have that third event whose relations to the others are unclear, namely the killing itself.

It is natural to assume that the action whose mention includes mention of an outcome itself somehow includes that outcome. Thus Feinberg says that a man's action may be 'squeezed down to a minimum or else stretched out' by the accordion effect. 'He turned the key, opened the door, he startled Smith, he killed Smith—all of these are things we might say that Jones *did* with one identical set of bodily movements', Feinberg tells us. It is just this relation of 'doing with' or 'doing by' in which we are interested. Feinberg continues: 'We can, if we wish, puff out an action to include an effect'.[11] Puffing

---

[11] Feinberg, 'Action and Responsibility', 146. I am concerned with an issue that is not central in Feinberg's excellent paper. Even if my caveats are justified, his thesis is not seriously affected.

out, squeezing down, stretching out sound like operations performed on one and the same event; yet if, as seems clear, these
operations change the time span of the event, then it cannot be one
and the same event: on Feinberg's theory, the action of opening the
door cannot be identical with the action of startling Smith. That this
is Feinberg's view comes out more clearly in the distinction he
makes between simple and causally complex acts. Simple acts are
those which require us to do nothing else (we have been calling
these primitive actions); causally complex acts, such as opening or
shutting a door, or startling, or killing someone, require us to do
something else first, as a means.[12] Thus Feinberg says, 'In order to
open a door, we must first do something else which will *cause* the
door to open; but to move one's finger one simply moves it—no
prior causal activity is required'.[13] He also talks of 'causally connected sequences of acts'.

The idea that opening a door requires prior causal activity, a
movement that caused the door to open, is not Feinberg's alone. He
quotes J. L. Austin in the same vein: '. . . a single term descriptive of
what he did may be made to cover either a smaller or a larger stretch
of events, those excluded by the narrower description being then
called "consequences" or "results" or "effects" or the like of his
act'.[14] Arthur Danto has drawn the distinction, in several articles,
between 'basic acts', such as moving a hand, and other acts that are
caused by the basic acts, such as moving a stone.[15]

It seems to me that this conception of actions and their consequences contains several closely related but quite fundamental confusions. It is a mistake to think that when I close the door of my own
free will *anyone* normally causes me to do it, even myself, or that
any prior or other action of mine causes me to close the door. So the
second error is to confuse what my action of moving my hand does
cause—the closing of the door—with something utterly different—my action of closing the door. And the third mistake, which is
forced by the others, is to suppose that when I close the door by
moving my hand, I perform two numerically distinct actions (as I
would have to if one were needed to cause the other). In the rest of
this paper I develop these points.[16]

---

[12] Ibid., 145.   [13] Ibid., 147.   [14] Austin, 'A Plea for Excuses', 145.
[15] Arthur Danto, 'What We Can Do', 'Basic Actions', 'Freedom and Forbearance'. Chisholm endorses the distinction in 'Freedom and Action', 39.
[16] Danto's view that if I close the door by moving my hand, my action of closing the

There is more than a hint of conflict between two incompatible ideas in Austin and Feinberg. As we noticed before, Feinberg shows some inclination to treat moving one's hand and opening the door (and startling Smith, etc.) as one and the same action, which is somehow stretched out or contracted; but he also says things that seem to contradict this, especially when he claims that one must first do something else to cause the door to open in order to open the door. The same strain is noticeable in Austin's pronouncement, for he speaks of different terms descriptive of *what the man did*—apparently one and the same thing—but the terms 'cover' smaller or larger stretches of events. Events that cover different stretches cannot be identical.[17]

There are, I think, insuperable difficulties that stand in the way of considering these various actions, the primitive actions like moving a hand, and the actions in describing which we refer to the consequences, as numerically distinct.

It is evident that the relation between the queen's moving her hand in such a way as to pour poison in the king's ear, and her killing him, cannot be the relation of event causality. If it were, we would have to say the queen caused herself to kill the king. This is not the same as saying the queen brought it about, or made it the case, that she killed the king; these locutions, while strained, do not seem clearly wrong, for it is not clear that they mean anything more than that the queen brought herself to kill the king. But then the locutions cannot be causal in the required sense. For suppose that by moving her hand the queen caused herself to kill the king. Then we could ask how she did this causing. The only answer I can imagine is that she did it by moving her hand in that way. But this movement was by itself enough to cause the death of the king—there was no point to a further action on the part of the queen. Nor is there any reason (unless we add to the story in an irrelevant way) why the queen should have wanted to cause herself to kill the king. What she wanted to do was kill the king—that is, do something that would cause his death. Is it not absurd to suppose that, after the queen has

---

door is caused by my moving my hand, has been ably criticized by Myles Brand, 'Danto on Basic Actions'; Frederick Stoutland, 'Basic Actions and Causality'; Wilfrid Sellars, 'Metaphysics and the Concept of a Person'.

My target is more general: I want to oppose any view that implies that if I do *A* by doing *B* then my doing *A* and my doing *B* must be numerically distinct.

[17] There is further discussion of these issues in Essays 6 and 8.

moved her hand in such a way as to cause the king's death, any deed remains for her to do or to complete? She has done her work; it only remains for the poison to do its.

It will not help to think of killing as an action that begins when the movement of the hand takes place but ends later. For once again, when we inquire into the relation between these events, the answer must be that the killing consists of the hand movement and one of its consequences. We can put them together this way because the movement of the hand caused the death. But then, in moving her hand, the queen was doing something that caused the death of the king. These are two descriptions of the same event—the queen moved her hand in that way; she did something that caused the death of the king. (Or to put it, as I would rather, in terms of a definite description: The moving of her hand by the queen on that occasion was identical with her doing something that caused the death of the king.) Doing something that causes a death is identical with causing a death. But there is no distinction to be made between causing the death of a person and killing him.[18] It follows that what we thought was a more attenuated event—the killing—took no more time, and did not differ from, the movement of the hand.

The idea that under the assumed circumstances killing a person differs from moving one's hand in a certain way springs from a confusion between a feature of the description of an event and a feature of the event itself. The mistake consists in thinking that when the description of an event is made to include reference to a consequence, then the consequence itself is included in the described event. The accordion, which remains the same through the squeezing and stretching, is the action; the changes are in aspects described, or descriptions of the event. There are, in fact, a great many tunes we can play on the accordion. We could start with, 'The queen moved her hand' and pull to the right by adding, 'thus causing the vial to empty into the king's ear'; and now another tug, 'thus causing the poison to enter the body of the king'; and finally (if we have had enough—for the possibilities for expansion are without clear limit), 'thus causing the king to die'. This expression can be shortened in many ways, into the centre, the left, or the right components, or any combination. For some examples: 'The queen

[18] See footnote 10. The argument goes through if the claim of this sentence is weakened by adding 'in the case where a person is killed by doing something that causes his death'.

moved her hand thus causing the death of the king' (the two ends); or, 'The queen killed the king' (collapse to the right); or, 'The queen emptied the vial into the king's ear' (the centre). There is another way to pull the instrument out, too: we could *start* with, 'The queen killed the king', adding 'by pouring poison in his ear', and so on—addition to the left. Many of these expressions are equivalent: for example, 'The queen killed the king by pouring poison in his ear' and, 'The queen poured poison in the king's ear thus causing his death'. And obviously the longer descriptions entail many of the shorter ones.

But this welter of related descriptions corresponds to a single descriptum—this is the conclusion on which our considerations all converge.[19] When we infer that he stopped his car from the fact that by pressing a pedal a man caused his automobile to come to a stop, we do not transfer agency from one event to another, or infer that the man was agent not only of one action but of two. We may indeed extend responsibility or liability for an action to responsibility or liability for its consequences, but this we do, not by saddling the agent with a new action, but by pointing out that his original action had those results.

We must conclude, perhaps with a shock of surprise, that our primitive actions, the ones we do not do by doing something else, mere movements of the body—these are all the actions there are. We never do more than move our bodies: the rest is up to nature.[20]

This doctrine, while not quite as bad as the bad old doctrine that all we ever do is will things to happen, or set ourselves to act, may seem to share some of the same disadvantages. Let me briefly indicate why I do not think that this is so.

First, it will be said that some actions require that we do others in order to bring them off, and so cannot be primitive: for example, before I can hit the bull's eye, I must load and raise my gun, then aim and pull the trigger. Of course I do not deny we must prepare the way for some actions by performing others. The criticism holds only if this shows some actions are not primitive. In the present example, the challenge is to demonstrate that hitting the bull's eye is a

---

[19] This conclusion is not new. It was clearly stated by G. E. M. Anscombe, *Intention*, 37–47. I followed suit in Essay 1.

[20] While not false, this sentence taken out of context is misleading. If I move the earth, this *sounds* like more than moving my body. The argument shows it is not. [Note added 1979.]

primitive action. And this it is, according to the argument I have given; for hitting the bull's eye is no more than doing something that causes the bull's eye to be hit, and this, given the right conditions, including a weapon in hand, I can do by holding my arms in a certain position and moving my trigger finger.

Second, it is often said that primitive actions are distinguished by the fact that we know, perhaps without need of observation or evidence, that we are performing them, while this is not a feature of such further events as hitting a bull's eye. But of course we can know that a certain event is taking place when it is described in one way and not know that it is taking place when described in another. Even when we are doing something intentionally, we may not know that we are doing it; this is even more obviously apt to be true of actions when they are described in terms of their unintended begettings.

Finally, it may seem a difficulty that primitive actions do not accommodate the concept of trying, for primitive actions are ones we just do—nothing can stand in the way, so to speak. But surely, the critic will say, there are some things we must strive to do (like hit the bull's eye). Once more the same sort of answer serves. Trying to do one thing may be simply doing another. I try to turn on the light by flicking the switch, but I simply flick the switch. Or perhaps even that is, on occasion, an attempt. Still, the attempt consists of something I can do without trying; just move my hand, perhaps.

The same fact underlies the last two answers: being attempted and being known to occur are not characteristics of events, but of events as described or conceived in one way or another. It is this fact too that explains why we may be limited, in our actions, to mere movements of our bodies, and yet may be capable, for better or for worse, of building dams, stemming floods, murdering one another, or, from time to time, hitting the bull's eye.

We may now return to the question of the relation between an agent and his action. The negative conclusion we have reached is this: the notion of cause has nothing directly to do with this relation. Knowledge that an action a has a certain upshot allows us to describe the agent as the cause of that upshot, but this is merely a convenient way of redescribing a, and of *it*, as we have seen, there is no point in saying that he is the cause. Causality allows us to redescribe actions in ways in which we cannot redescribe other events; this fact is a mark of actions, but yields no analysis of agency.

To say that all actions are primitive actions is merely to acknowledge, perhaps in a misleading way, the fact that the concept of being primitive, like the concept of being intentional, is intensional, and so cannot mark out a *class* of actions. If an event is an action, then under some description(s) it is primitive, and under some description(s) it is intentional. This explains why we were frustrated in the attempt to assume a basic concept of agency as applied to primitive actions and extend it to further actions defined in terms of the consequences of primitive actions: the attempt fails because there are no further actions, only further descriptions.

The collapse of all actions into the primitive, which is marked in syntax by the accordion effect, leads to a vast simplification of the problem of agency, for it shows that there is a relation between a person and an event, when it is his action, that is independent of how the terms of the relation are described. On the other hand, we have discovered no analysis of this relation that does not appeal to the concept of intention. Whether intention can be understood in terms of more basic or simple ideas is a question with which I have not tried to cope in this paper.

4 *Freedom to Act*

The view that free actions are caused by states and episodes like desires, beliefs, rememberings, and the promptings of passion comes under fire from two quarters. There are the broadsides from those who believe they can see, or even prove, that freedom is inconsistent with the assumption that actions are causally determined, at least if the causes can be traced back to events outside the agent. I shall not be directly concerned with such arguments, since I know of none that is more than superficially plausible. Hobbes, Locke, Hume, Moore, Schlick, Ayer, Stevenson, and a host of others have done what can be done, or ought ever to have been needed, to remove the confusions that can make determinism seem to frustrate freedom.

The other attack is more interesting. It is aimed not at determinism as such but at the causal theory of action. If a free action is one that is caused in certain ways, then freedom to act must be a *causal power* of the actor that comes into play when certain conditions are satisfied. The champion of the causal theory cannot evade the challenge to produce an analysis of freedom to act which makes it out to be a causal power, or so at least it seems. Here the causal theorist has felt forced into a defensive stance, for no proposed account meets all objections. What is worse, the faults in known analyses show recurring patterns, suggesting the impossibility of a satisfactory solution. In brief outline, here is one familiar and disturbing difficulty.

It is natural to say that a person can do something (or is free to do it) if all that is required, if he is to do it, is that he *will* to do it. But does 'He would do it if he were to will to do it' express a causal disposition? If willing is an act distinct from doing, it might be a cause, but the question would then arise when an agent is free to will. If willing is not an act distinct from doing, then it cannot be a cause of the doing. The dilemma is not resolved by substituting choosing, deciding, intending or trying for willing; most of these alternatives merely make it more obvious which horn does the impaling. If this difficulty cannot be overcome, shouldn't we abandon the causal theory of action, even if it doesn't fall to frontal assault?

The particular dilemma just mentioned can, I believe, be resolved. Many other usual objections to a causal analysis of freedom can also be circumvented. A central difficulty nevertheless remains which permanently frustrates the formulation of a satisfactory analysis; but this fact ought not to persuade us that freedom to act is not a causal power. Broadly adumbrated, this is the view for which I shall now argue. First, something more must be said about the idea of a causal power.

By a causal power, I mean a property of an object such that a change of a certain sort in the object causes an event of another sort. This characterization is, of course, much in need of clarification. The appeal to sorts of change, for example, invites discussion of causal laws, counterfactuals, and natural kinds. I hope to avoid the worst troubles that lurk here, however, by holding assumptions down. In particular, I shall not take it as given that causal powers can be analysed by subjunctive conditionals, though much of the discussion will concern the propriety of proposed subjunctive conditional analyses.

The concept of a causal power is indifferent to the intuitive distinction between the active and passive. We may think of solubility as a passive characteristic, and of being a solvent as active. Yet they are logically on a par, the former being a property of things changes in which cause them to dissolve, and the latter being a property of things changes in which cause other things to dissolve. Both are equally causal powers.

Many attacks on the causal theory of free action fasten on powers like solubility as if only such powers could provide analogues for being free to act. This puts an unwarranted strain on the theory. If a substance is soluble, certain eventualities cause it to dissolve; analogy suggests that a free action is one where the agent is caused to act by something that happens to him. Yet an equally good analogy would say that a free action is one where a change in the agent causes something to happen outside him. This would, in fact, be the natural meaning of 'causal power' attributed to a person: he has the power to cause things to happen. Again, however, the details of the parallel are wrong. It is true that a person is free to bring about whatever would be brought about by actions he is free to perform. But this concept of what an agent is free to bring about depends on a prior idea of what he is free to do. A man might have the power to destroy the world, in the sense that if he were simply to push a button before him, the world would be destroyed. Yet he would not be free to destroy the world unless he were free to push the button.

Another kind of causal power is such that an object that possesses it is caused to change in a certain way if a prior change takes place in the object. An example might be the property of a tank that is self-sealing. If the tank starts to leak, this causes the leak to seal. Similar properties are: being in equilibrium, unstable, homeostatic. If freedom to act is a causal power, it belongs to this category.

There is little difference, perhaps, between 'The heat caused the flower to wilt' and 'The heat caused the wilting of the flower.' But there can be a great difference between 'The heat caused Samantha to return to Patna' and 'The heat caused Samantha's return to Patna.' The former implies, or strongly suggests, a limitation on Samantha's freedom of action; the latter does not. In testing the plausibility of the claim that freedom to act is a causal power we should therefore bear in mind the neutral reading. An *action* may be caused without the *agent* being caused to perform it. Even when the cause is internal, we sense a difference. 'Desire caused him to do it' suggests a lack of control that might be called on to excuse the action because it makes the act less than voluntary; 'Desire was the cause of his doing it' leaves the question of freedom open.

Turning from these preliminary matters, let us consider arguments designed to show that being free to perform an action cannot be a causal power. We may begin with some of the claims in J. L. Austin's 'Ifs and Cans'.[1] According to Austin, G. E. Moore suggested the following three propositions, the first two explicitly, and the third implicitly:

(1) 'Could have' simply means 'could have if I had chosen'.
(2) For 'could have if I had chosen' we may substitute 'should have if I had chosen'.

---

[1] 'Ifs and Cans'.

(3) The *if*-clauses in these expressions state the causal conditions upon which it would have followed that I could or should have done the thing different from what I did actually do.

Austin argues that (1) to (3) are false. I shall argue that for all Austin says they may be true (though there is an important point on which Austin is right: one of the disjuncts of (3) is false).

Let's suppose that Moore's basic view was that to say an agent could have done something, or can now do something (that is, that he was, or is, free to do it) means that if he had chosen to do it, he would have done it (or if he now chooses to do it, he does or will do it). I have transposed the thesis from the first to the third person in order to forestall irrelevant qualms over performances, and I have generalized a bit on the tenses. (Difficulties over tense and time remain, but I do not believe we need to resolve them here.) To settle on one formulation, take Moore's thesis to be: '*A* can ——' means 'If *A* chooses to ——, then he ——s.' (The 's' at the end is to be added to the main verb, even if it does not conclude the phrase substituted for the blank. Thus, 'Jones can anchor his boat in the channel' means 'If Jones chooses to anchor his boat in the channel, then he anchors his boat in the channel.') I shall examine Austin's attack on Moore in the light of this proposal.

If we accept the suggestion, then '*A* can ——' does mean the same as (or anyway is logically equivalent to) '*A* can —— if he chooses.' For given Moore's suggestion, '*A* can ——' means '*A* ——s if he chooses', while '*A* can —— if he chooses' expands into '*A* ——s if he chooses, if he chooses', and these last two sentences are logically equivalent.

Moore sees no important difference between '*A* ——s if he chooses' and '*A* can —— if he chooses.' And given his suggested analysis of 'can', he is surely right, for the second sentence becomes, as we just saw, '*A* ——s if he chooses, if he chooses', which is logically equivalent to '*A* ——s if he chooses.' Austin argues against this equivalence, but the argument is faulty. He assumes as obvious that if '*A* ——s if he chooses' means the same as '*A* can —— if he chooses', then '——s' and 'can ——' must mean the same; and this is patently false. But the assumption is wrong. '——s' and 'can ——' don't mean the same, and yet the sentences '*A* ——s if he chooses' and '*A* can —— if he chooses' are equivalent—

granted Moore's suggestion that 'can —' means '——s if he chooses'.[2]

Austin emphasizes the fact that not all *if* clauses introduce causal conditions; some sentences in which 'if' is the main connective are causal conditionals and some are not. Austin is obviously right in saying that '*A* can —— if he chooses' is not a causal conditional, at least given Moore's analysis of 'can'. The reason is that '*A* can ——' is itself (we are supposing) a conditional, the antecedent of which is 'He chooses'. So '*A* can —— if he chooses' logically entails the consequent, which shows that the longer sentence can't be a causal conditional. It would be a blunder, however, to reason that since '*A* can —— if he chooses' is logically equivalent to '*A* ——s if he chooses' and the former is not a causal conditional, therefore the latter is not a causal conditional.

Austin seems to have thought that the word 'if' has a different meaning in causal conditionals than elsewhere, but the evidence for this is not very convincing. There is no good reason to say that 'if' has different meanings in 'If a number is divisible by four, it is divisible by two' and 'If you eat enough apple seeds you will get arsenic poisoning', though the latter is presumably a causal conditional and the former is not. The difference is not semantical but epistemological: the first sentence we hold to be true for logical or arithmetic reasons; the second sentence we would hold to be true only if we thought eating apple seeds *caused* arsenic poisoning.

The question of meaning is relevant to the next point. Austin says it is characteristic of causal conditionals that contraposition is a valid form of inference. This is surely right, and we can see that contraposition does work for '*A* ——s if he chooses'. But are there any *if*-sentences for which contraposition does not work? (If there are, we must admit that 'if' is ambiguous.) I do not think so. We are not inclined to *reason* from '*A* can —— if he chooses' to its contrapositive, but the explanation is not, I think, that the formal entailment fails. A better explanation is this: on Moore's analysis, '*A* can —— if he chooses' means '*A* ——s if he chooses, if he chooses', which in turn is logically equivalent to its consequent. The question whether '*A* can —— if he chooses' entails its contraposi-

---

[2] In 'Ifs, Cans and Causes' Keith Lehrer notes the equivalence, on Moore's analysis, of 'I shall if I choose' and 'I can if I choose.' But then he copies Austin's *non sequitur*: 'I can, if I choose' is clearly not equivalent to 'I shall, if I choose' because "I can" is not equivalent to "I shall."' '

tive is therefore just the question whether '*A* can ——' entails 'If *A* can't ——, then he doesn't choose to ——.' Of course it does entail it, in the formal and vacuous way a contradiction entails anything at all. This is not the kind of entailment that invites serious reasoning outside of logic. Austin's argument helps us see, once more, that '*A* can —— if he chooses' isn't a causal conditional, but it doesn't show that a causal 'if' isn't in the offing—i.e. in the 'can'.[3]

In a review of Austin's papers,[4] Roderick Chisholm comes, by a somewhat different route, to essentially the same conclusion I have reached, that Austin has given no conclusive reason for rejecting Moore's analysis of 'can' in terms of 'does if he chooses', nor have the resulting sentences been shown not to be causal conditionals. Chisholm believes Austin has shown that 'He can —— if he chooses' is not a conditional, and that there are difficulties about 'choose' as the appropriate verb for the antecedent. I don't think Austin succeeds in demonstrating either of these things, but it doesn't matter so far as Chisholm's argument is concerned. For Chisholm's argument, if it is a good one, applies not only to 'choose', but also to many other verbs we might try in its place, like 'try'.

Chisholm's point is that while it might be true that if a man were to choose, or try, to perform some action, then he would perform that action, nevertheless he might be unable to choose or to try, in which case, he couldn't perform the action. So it might be true that a person would —— if he tried, yet false that he could ——.

I think Chisholm's argument is very important, and that he is right in saying that to overlook it is to make a 'first-water, ground-floor mistake' (Austin's words). What the argument shows is that the antecedent of a causal conditional that attempts to analyse 'can' or 'could' or 'free to' must not contain, as its dominant verb, a verb of action, or any verb which makes sense of the question, Can someone do *it*?

If I am not mistaken, M. R. Ayers has supposed that because the

---

[3] For analogous reasons, I would maintain that 'There are biscuits on the sideboard if you want them' contains a normal 'if' and does entail its contrapositive. We miss this because we assume that mere wanting can't produce biscuits; therefore, 'There are biscuits on the sideboard if you want them' is true only if there are biscuits on the sideboard. This assumption contradicts the antecedent of the contrapositive, so instead of inferring the contrapositive, we are simply baffled by the question whether it is true.

[4] 'J. L. Austin's *Philosophical Papers*', 20–5.

antecedent of a causal conditional analysing freedom to act cannot be dominated by a verb of action, no causal analysis is possible. His argument seems to be that in all other cases of causal conditionals, the antecedent is something we can make true independently of the supposed effect. But in the case of action, he says, 'that testing one's ability to do something is not the same kind of thing as testing the truth of a hypothesis, simply follows from the truism that in order to do something it is not always necessary to do something else first'. He concludes, 'Thus the possibility of any hypothetical analysis of personal power is ruled out.'[5]

The conclusion does not follow from the premise. What follows from the truism that 'in order to do something it is not always necessary to do something else first' is only that if a causal conditional is to analyse freedom to act, the antecedent cannot have as its main verb a verb of action. Of course, this does mean the agent himself cannot normally test the truth of the conditional by doing something else—by making such an antecedent true. It does not follow, however, that he cannot test in some other way whether he has the ability to do something, nor does it follow that ability may not consist in a causal power. All we can conclude is that the test cannot (always, anyway) be carried out by the agent's doing something to create the conditions for actualizing the power.

Keith Lehrer has urged, on far more general grounds, that no causal conditional can give the meaning of 'A is free to —— (or can ——).' He claims that 'A——s if condition C obtains' is consistent with the following two statements: 'A cannot —— if condition C does not obtain' and 'C does not obtain.' Yet if these two last sentences *are* true A cannot ——. In Lehrer's words, 'it is logically possible that some condition which is a sufficient condition to cause a person to do something should also be a necessary condition of his being able to do it, and that the condition should fail to occur'.[6]

Although Lehrer does not seem to have noticed it, his argument has nothing in particular to do with action. If his line of reasoning is sound, it shows that no attribution of a power or disposition is ever equivalent to a conditional (whether the conditional is construed as causal law or subjunctive). Thus on Lehrer's argument, to say something is water-soluble can't mean it dissolves if placed in water, since it may not be soluble and yet placing it in water would make it dissolve.[7] There are various ways one may try to cope with this point. But for my purposes it will be enough to remark that even if Lehrer's argument showed that no disposition or power is correctly analysed by a subjunctive conditional, the claim that being able to do something is to have a causal power would not be undermined.

There are, of course, cases where wanting, trying, choosing, or even intending, to do something prevents us from doing that very thing, even though there may be a sense in which we could have done it (if we hadn't wanted to, tried to, etc.). Moralists from Aristotle to Mill have pointed out that trying to be happy is unlikely to produce happiness, and Schlick was so convinced of this that he revised hedonism to read, not, 'Do what you can to be happy', but 'Be ready for happiness'. Austin gives a more pedestrian example: a man may try to (choose to, want to, intend to) sink a putt and fail, while not admitting that he couldn't. None of these cases is like Lehrer's. Rather they are cases where one can, and yet although one tries (chooses, intends, wants), one doesn't bring it off.

Chisholm proposes a way around this difficulty. He suggests that *A* can do something provided there is something (different or the same) such that if *A* tries (chooses, wants, intends) to do *it*, then he does do the first thing. In the case of the putt, this means the agent can sink the putt if there is something else, perhaps hit the ball an inch to the left of the cup, such that if he tries to do *it*, then he will sink the putt.

Neither Chisholm nor I thinks this is the correct analysis, only that it answers Austin's objection to Moore's proposal. But there is another fact (not given by Chisholm) that recommends Chisholm's revision of Moore. When we describe our actions, we include not only what we do intentionally but also many things we do unintentionally. If one holds, as I do, that unintentional actions are intentional actions under other descriptions, then the point may be put by saying that action-descriptions include descriptions of actions as intentional, and certain other descriptions of those same events. For example, I may intentionally put down a coin on the counter of a newsstand in Paris, intending to pay for a purchase. Unintentionally, I put down a drachma instead of a franc, thus provoking some

[7] I am grateful to G. E. M. Anscombe for correcting my interpretation of Lehrer's argument. The correct response to Lehrer is simply that if one analyses solubility by a causal conditional, one can't consistently allow that what causes dissolving is also a necessary condition of solubility, since in that case the only soluble things would be dissolved. Similarly for conditional analyses of freedom. [Footnote added 1977.]

interesting argot. My one action can be correctly characterized as: putting down a coin, putting down a drachma, provoking some interesting argot. Only under the first characterization is my act intentional.

An important way of describing actions (and other events) is in terms of their effects or consequences. Provoking some interesting argot characterizes my act of putting down the coin in terms of a consequence, namely the provoked response. The provoked response is not, of course, *part* of my action, only part of its description. We may call the effect caused by my action the *completion* of the action. My actions include, then, what I do intentionally, and anything I do the completion of which is caused by an action. The second clause adds no new events, only new ways of describing the old.[8]

The analysis of the concept of an action comes, then, if I am right, in two stages, the first bringing in intention directly, and the second extending the concept to actions in the first sense, redescribed in terms of unintended consequences or other unintended characteristics. Since what we do do is surely included in what we can do, the pattern of analysis that applies to the concept of action ought to work for the notion of what an agent can, or is free to, do. He is free to do those things he would do if he chose (wanted, tried) to do them; but he is also free to do those things whose completions would be caused by an action he is free to do in the first sense.

This is a very generous way of counting the things we can do—it includes all the things we could bring about through our intentional action, whether by plan or by accident, through blind luck, or masterful contrivance. In this broad sense of 'can' or 'could', every one of us could make a million dollars on the stock market, marry a movie star, or even bring an end to the war. If only we knew how.

For practical purposes we are often interested in more limited concepts of what we can do—what we do do if we want or intend or try to do *it*, for example, or what we can do reliably. But the broader concept I have introduced is basic, at least in the sense that the others could be defined in terms of it—assuming, of course, that we have things right so far. It is simply a more inclusive concept than we usually put to work, because it includes in the things we can do things we would do only by luck or if we knew more than we do. But it is not so broad as to include the merely possible, logically or physically—what someone could do for all that nature cares. This

[8] See also Essay 3.

category would embrace things I could not do intentionally, and that would not be caused by any intentional act I could perform, like run a mile in four minutes.

We have now decided on a two-part theory of what a man can do: one part aims to explain what he can do intentionally, and the other part extends this to what the intentional would cause. I want to concentrate on the first part, and for the moment on the main verb of the antecedent. To say a man can do *x* intentionally is to say he would do it if—what?

The problem is the one we sketched at the beginning. In order to be eligible as a cause, the event mentioned must be separate from the action; but if it is separate from the action, there is, it seems, always the possibility of asking about *it*, whether the agent is free to do it. The objection applies to choosing, willing, intending, and trying. None of these is plausibly the cause of an action, because normally these are ways of characterizing the action chosen, willed, intended, or tried, not descriptions of further actions, events, or states. Thus if a man tries to hit a home run and succeeds, his try *is* his success, and cannot be its cause. Sometimes choosing or willing to do something are mental acts performed prior to doing the things but in these cases the question does arise whether the agent is free to do them. The analysis fails for another reason, for a man may choose or decide on Monday to swallow a button on Tuesday and yet fail to do it, and this would not necessarily show that he could not swallow a button on either Monday or Tuesday—or even that he could not, on Monday, swallow a button on Tuesday. Finally, we should remember that a person may be able to do something, and may actually do it intentionally, and yet never have chosen, decided, tried, or intended to do it.

The only hope for the causal analysis is to find states or events which are causal conditions of intentional actions, but which are not themselves actions or events about which the question whether the agent can perform them can intelligibly be raised. The most eligible such states or events are the beliefs and desires of an agent that *rationalize* an action, in the sense that their propositional expressions put the action in a favourable light, provide an account of the reasons the agent had in acting, and allow us to reconstruct the intention with which he acted. For example, suppose a man saws a piano in half because he wants to throw the piano out the window and he believes that sawing the piano in half will promote his enterprise. His reason for sawing the piano in half was that he wanted to throw the piano out the window—to which we may add that he believed he could do this if he first sawed the piano in half. Obviously his reasons, or the beliefs and desires that correspond to them, explain why he acted as he did. And we can see that the intention with which he sawed the piano in half was to get the piano out the window.

So now I suggest that we consider the following formulation of the first part of our thoery of what a man can do:

A can do *x* intentionally (under the description *d*) means that if *A* has desires and beliefs that rationalize *x* (under *d*), then *A* does *x*.

A number of previous problems seem to be solved by this analysis. The antecedent condition (*A* has desires and beliefs that rationalize *x*) is prior to and separate from the action, and so is suited to be a cause (in this case, it is a state rather than an event—but this could be changed along these lines: 'coming to have desires and beliefs that rationalize *x*'). The antecedent condition does not mention something that is an action, so the question whether the agent can do it does not arise. The point isn't that desires and beliefs aren't ever in an agent's control, but rather that coming to have them isn't something an agent does. I do not want to suggest that the nature of an agent's beliefs and desires, and the question how he acquired them, are irrelevant to questions of how free he, or his actions, are. But these questions are on a different and more sophisticated level from that of our present discussion.[9]

By adding 'intentionally' to the definiendum, we have answered one of Austin's objections to a conditional analysis of 'can'. For one of his objections to 'I shall if I choose' was that the consequent alone

[9] Not everyone agrees that it helps solve our problem to distinguish between antecedent conditions that are actions and antecedent conditions that are not. Thus Chisholm says conditional analyses of 'can do ——' are questionable even if the antecedent of the analysans does not concern an action (op. cit., 25). In connection with the theory under consideration, therefore, Chisholm would say that the agent is not free to —— even if he would —— given that he had appropriate desires and beliefs, provided he could not have the desire or belief. In this essay I reject this view. I hold that there is a basic sense in which we are free to —— (can ——) provided all that is needed for the action (in addition to other present circumstances) is the right attitudes and beliefs. The question whether we can have those attitudes and beliefs in turn is (except in special circumstances) not relevant. Some hold that to admit this is at least to allow that 'can' has a different meaning in 'can ——' (meaning 'is free to ——') and in 'can believe ...' or 'can want ...'. I do not see that this conclusion follows, but perhaps it is true. [Footnote added in 1979.]

entails 'I can.' Similarly, if we were to analyse '*A* can do *x*' as 'If *A* has attitudes that rationalize *x*, then *A* does *x*', Austin would doubtless complain that '*A* does *x*' again entails '*A* can do *x*', showing that there is something odd about the conditional. The difficulty disappears when the definiendum is '*A* can do *x* intentionally', for this *doesn't* follow from '*A* does *x*.' But more important, the reason for the change shows that the objection was not valid in the first place. In fact, doing *x* *doesn't* entail that one can do *x*.[10] The word 'doing' suggests intention somewhere, but in the present context, it serves only as a blank to be filled by any verb whose subject can name or describe a person. It doesn't follow, from the fact that at noon today I became exactly 55 years, 72 days, 11 hours, and 59 minutes old, that I was free to become this age (rather than another) on this date, or that I could do it. It simply wasn't in my power to do it—no desire or belief of mine played any causal role in my becoming this age on this date, nor could it have. If, on the other hand, rationalizing attitudes do cause an action of mine, then not only does the action occur, but it is, under the rationalized description, intentional. This is, of course, what we should expect: what an agent does do intentionally is what he is free to do *and* has adequate reasons for doing.

The discussion of the last few paragraphs may help to put another familiar puzzle in perspective. It is natural to suppose that an action that one is free to perform is an action that one is also free *not* to perform. Similarly, we find many philosophers maintaining that an action is freely performed only if the agent was, when he did it, free to abstain: he could have done otherwise. For suppose he could not have done otherwise; doesn't this mean he had no choice? If the causal theory drives us to this, why isn't the libertarian right when he denies that freedom can be reconciled with the causal theory? Or to take the merely potential case, isn't it an empty pretence to say a man is free to perform an action if he is not also free not to perform it? Surely, freedom means the existence of alternatives.

The difficulty (recently brought to the fore by Harry Frankfurt[11]) is that if we say a man is free to do *x* only if his doing *x* depends on whether or not the attitudinal condition holds (he chooses to do *x*, decides to, wills it, has rationalizing attitudes), then we find counter

---

[10] David Pears makes the same observation, though for somewhat different reasons, in 'Ifs and Cans', 382.
[11] 'Alternate Possibilities and Moral Responsibility'.

instances in cases of overdetermination. What a man does of his own free will—an action done by choice and with intent, caused by his own wants and beliefs—may be something he would have been caused to do in another way if the choice or motives had been lacking.

Two intuitions seem at war, and the territory that is threatened with destruction is occupied by the causal theory. The intuitions are, on the one hand, the view that we cannot be free to do what we would be causally determined to do in any case, and on the other hand, the feeling that if we choose to do something and do it because we chose it, then the action is free no matter what would have happened if we *hadn't* chosen.

The puzzle is resolved by a discovery we made in another context. What depends on the agent is the intentional performance of an act of a certain sort. It is true that it may sometimes be the case that what a man does intentionally he might have been caused to do anyway, by alien forces. But in that case what he would have done would not have been intentional. So even in the overdetermined cases, something rests with the agent. Not, as it happens, *what* he does (when described in a way that leaves open whether it was intentional), but whether he does it intentionally. His action, in the sense in which action depends on intentionality, occurs or not as he wills; what he does, in the broader sense, may occur whether or not he wills it.

These reflections show, I think, that it is an error to suppose we add anything to the analysis of freedom when we say an agent is free to do something if he can do it *or not*, as he pleases (chooses, etc.). For if this means that he acts freely only if he would not have done what he did had the attitudinal conditions been absent, it is false. And if it means he would not have acted intentionally had the attitudinal conditions been absent, then this is true, but it is gratuitous to say it since it is a logical truth.

I have been examining the concept of what a person can do, or is free to do, when this concept is interpreted to include what he can do, whether he knows it or not, by doing something intentional. On the present proposal, a man is free to assassinate a future tyrant if he is standing next to the future tyrant, has a gun, and otherwise the opportunity, even if he has no reason for thinking the man next to him is the future tyrant. For *if* he believed the man next to him was the future tyrant and he wanted above all to kill the future tyrant, then, given the propitious circumstances, he would kill him. We get a more limited version of what he can do if we say he can kill the future tyrant only if, given that he wants to kill the future tyrant, he does. For then, since he will not do it intentionally unless he knows or believes the man next him is the future tyrant, he is not free to do it unless he has the knowledge. Obviously this more restricted sense of being able to do something is important when it comes to holding a man responsible for failing to act.

My purpose in pointing out how we can adjust the conditions under which a man acts to accommodate various interpretations of freedom is to make plain the extent to which the problem of analysing freedom to act is interlocked with the problem of defining intentional action (and with it, action in general, since non-intentional actions are intentional actions under other descriptions). The reason is obvious: to say when an agent is free to perform an action intentionally (i.e. with a certain intention) is to state conditions under which he would perform the action; to explain the performance of an action with a certain intention is to say that the conditions are satisfied. Of course, this style of analysis and explanation works only if the satisfaction of the conditions (which we are assuming are causal) always leads to the performance of the action. And so, if we can analyse either freedom to act (in our broad sense) *or* acting with an intention, we will be in command of a law saying that whenever certain conditions are satisfied, an agent will perform an action with a certain intention.

Since intentional action is, on the causal theory, partly defined by its causes, it should perhaps not surprise us that merely knowing the analysis of intentional action would put us in command of a law of behaviour. Still, how can what seems to be an empirical generalization emerge from wholly analytic considerations? I shall return to this question presently.

These things, then, stand together: a law stating conditions under which agents perform intentional actions; an analysis of freedom to act that makes it a causal power; and a causal analysis of intentional action. They stand together, or they fall together. In my opinion, they fall together if what we want are explicit, non-questionbegging analyses, or laws without generous caveats and *ceteris paribus* clauses.

A brief discussion of three recent attempts to provide the appropriate law or analysis will help me explain the difficulties.

First consider the following law of action proposed by Paul Churchland:[12]

> *If* A wants ø and believes x-ing is a way to bring about ø and that there is no better way to bring about ø, and A has no overriding want, and knows how to x, and is able to x, *then* A x's. [I have simplified Churchland's formulation.]

Clearly if this proposal were acceptable we could define being able to x as the state of an agent who, when the other conditions are satisfied, x's. But the 'law' is defective. For one thing, the notion of 'overriding want' is treacherous, since it is unclear how a want is shown to be overriding except by the fact that it overrides. The more general difficulty, however, is this: all the conditions may be satisfied, and yet they may fail to ignite an action. It might happen as simply as this: the agent wants ø, and he believes x-ing is the best way to bring about ø, and yet he fails to put these two things together; the practical reasoning that would lead him to conclude that x is worth doing may simply fail to occur. There is no more reason to suppose that a person who has reasons for acting will always act on them than to suppose that a person who has beliefs which entail a certain conclusion will draw that conclusion.

Alvin Goldman illustrates the interlocking of the three enterprises I have been discussing in this way. First he defines a basic act-type as a kind of act an agent performs if no external forces prevent him and he wants to perform an act of that type. He goes on:[13]

if x is a basic act-type for A at t, and if A is in standard conditions with respect to x at t, then the following causal conditional is true: 'If A wanted to do x at t, then A would do x at t.' This causal conditional statement . . . entails the statement that A is 'able' to do x at t. [I have altered the letter symbolism.]

Because Goldman makes wanting to do x the attitudinal cause of doing x, his analysis may seem to get around our difficulty. But in fact it does not, as we see when we clearly distinguish act-types and particular actions. Wanting to do x is, for Goldman, a desire for a type of action, not for a particular action. This is fine: the objects of wanting and desiring are propositional in character, and so cannot be particular actions. But then it is a mistake to say that wanting to do x (i.e. to perform an action of type x) causes x, since x, not being

[12] 'The Logical Character of Action Explanations', 221–2.
[13] *A Theory of Human Action*, 199.

an event, but a type, cannot be caused. What can be caused is only some event that belongs to the type. Which one of these is to be caused? Goldman enunciates a law that, rephrased, would go like this: if a person is not restrained by external forces, then if he wants to do something of type x, he does something of type x. This is supposed to apply only to 'basic' acts, but in fact I do not believe it applies to any interesting class of actions. The trouble is that wanting to do something of type x, even if x-ing is completely in his power, and he knows it, may not cause an agent to do anything of type x. And a worse trouble is that wanting to do something of type x may cause someone to do something of type x, and yet the causal chain may operate in such a manner that the act is not intentional.

David Armstrong, in a paper called 'Acting and Trying',[14] comes closest to seeing the nature of the difficulty. He asks the question what we must add to 'A tried to do x' in order to have necessary and sufficient conditions for 'A did x intentionally.' Most of us would say—I certainly would—that trying itself isn't necessary in many cases, but this point, though at the heart of Armstrong's interest, is largely irrelevant to the present theme. According to Armstrong, A's doing x intentionally entails that A's trying to do x caused the occurrence of x, but A's trying to do x, even if it causes the occurrence of x does not prove that A did x intentionally. The difficulty is, that the attempt may bring about the desired effect in an unexpected or undesired way. Here is an example of Daniel Bennett's. A man may try to kill someone by shooting at him. Suppose the killer misses his victim by a mile, but the shot stampedes a herd of wild pigs that trample the intended victim to death. Do we want to say the man killed his victim *intentionally*? The point is that not just any causal connection between rationalizing attitudes and a wanted effect suffices to guarantee that producing the wanted effect was intentional. The causal chain must follow the right sort of route.

Armstrong tries to fill this gap by saying that the wanted effect must be produced by a causal chain that answers, at least roughly, to the pattern of practical reasoning. In Bennett's example, we must suppose the agent intended to kill the victim by pulling the trigger because he reasoned that pulling the trigger would cause the gun to fire, which would cause the bullet to fly, which would cause the bullet to penetrate the body of the victim, thus causing his death.

[14] 'Acting and Trying'. Armstrong has replied to my comments in a later issue of *Philosophical Papers* (May 1975).

But the pattern of events portrayed by the pattern of practical reasoning was not produced by the action of pulling the trigger. This throws in doubt the question whether the agent intentionally killed his victim.

I am not sure whether or not this difficulty can be overcome, but there is a related problem that Armstrong does not consider which seems to me insurmountable. This is the problem, not of quaint *external* causal chains, but of nonstandard or lunatic *internal* causal chains. Armstrong, in trying to mend the trouble about unwanted external causal relations, was pushed into talking of the course of practical reasoning, the way in which beliefs and desires interact to produce action. And here we see that Armstrong's analysis, like the one I proposed a few pages back, must cope with the question *how* beliefs and desires cause intentional actions. Beliefs and desires that would rationalize an action if they caused it in the *right* way—through a course of practical reasoning, as we might try saying—may cause it in other ways. If so, the action was not performed with the intention that we could have read off from the attitudes that caused it. What I despair of spelling out is the way in which attitudes must cause actions if they are to rationalize the action.

Let a single example serve. A climber might want to rid himself of the weight and danger of holding another man on a rope, and he might know that by loosening his hold on the rope he could rid himself of the weight and danger. This belief and want might so unnerve him as to cause him to loosen his hold, and yet it might be the case that he never *chose* to loosen his hold, nor did he do it intentionally. It will not help, I think, to add that the belief and the want must combine to cause him to want to loosen his hold, for there will remain the *two* questions *how* the belief and the want caused the second want, and *how* wanting to loosen his hold caused him to loosen his hold.

Some distance back, I tried analysing, 'A is free to do *x* (or can do *x*)' in terms of the conditional, 'He would do *x* intentionally if he had attitudes that rationalized his doing *x*'. Even if we read this subjunctive conditional as implying a causal relation, we can see now that it is not adequate. If the agent does *x* intentionally, then his doing *x* is caused by his attitudes that rationalize *x*. But since there may be wayward causal chains, we cannot say that if attitudes that would rationalize *x* cause an agent to do *x*, then he does *x* intentionally.

It is largely because we cannot see how to complete the statement of the causal conditions of intentional action that we cannot tell whether, if we got them right, the result would be a piece of analysis or an empirical law for predicting behaviour. If we begin with the suggestion that if a man has beliefs and desires that rationalize an action of type x, then he intentionally performs an action of type x, it is easy to see that we improve matters if we add to the antecedent 'and those beliefs and desires, or the coming to have them, cause the agent to perform an action of type x'. To improve on this formulation in turn, in a way that would eliminate wrong causal chains, would also eliminate the need to depend on the open appeal to causal relations. We would simply say, given these (specified) conditions, there always is an intentional action of a specified type. This would be understood as a causal law, of course, but it would not need to mention causality. Unavoidable mention of causality is a cloak for ignorance; we must appeal to the notion of cause when we lack detailed and accurate laws. In the analysis of action, mention of causality takes up some of the slack between analysis and science.

If the finished formulation of the proposed law were produced, we could try to judge its character more accurately. If the terms of the antecedent conditions were to remain mentalistic (for example by using such concepts as those of belief and desire), the law would continue to seem constitutive or analytic. But the attempt to be more accurate, to eliminate appeal to questionbegging notions like 'overriding' desires, would, I think, suggest that a serious law would have to state the antecedent conditions in physical, or at least behaviouristic, terms. Then the law would be more clearly empirical. But the attempt to find such a law would also end in frustration, though for different reasons.[15]

We must count our search for a causal analysis of 'A is free to do x' a failure. Does this show that Austin was, after all, right? Well, he was right in rejecting Moore's analysis (though many of his reasons were, I have argued, defective). And he was right, I believe, in holding that no other causal conditional could analyse 'A is free to do x.' But Austin also believed, however tentatively, that he had shown that being free to perform an action cannot be a causal power at all, and hence that the causal theory of action is false (he calls it 'determinism', but I don't want to go into that). On this last and

---

[15] I discuss why strict psycho-physical laws are impossible in Essays 11 and 12.

important point he was, I think, wrong. For although we cannot hope to define or analyse freedom to act in terms of concepts that fully identify the causal conditions of intentional action, there is no obstacle to the view that freedom to act is a causal power of the agent.

5 *Intending*

Someone may intend to build a squirrel house without having decided to do it, deliberated about it, formed an intention to do it, or reasoned about it. And despite his intention, he may never build a squirrel house, try to build a squirrel house, or do anything whatever with the intention of getting a squirrel house built. Pure intending of this kind, intending that may occur without practical reasoning, action, or consequence, poses a problem if we want to give an account of the concept of intention that does not invoke unanalysed episodes or attitudes like willing, mysterious acts of the will, or kinds of causation foreign to science.

When action is added to intention, for example when someone nails two boards together with the intention of building a squirrel house, then it may at first seem that the same problem does not necessarily arise. We are able to explain what goes on in such a case without assuming or postulating any odd or special events, episodes, attitudes or acts. Here is how we may explain it. Someone who acts with a certain intention acts for a reason; he has something in mind that he wants to promote or accomplish. A man who nails boards together with the intention of building a squirrel house must want to build a squirrel house, or think that he ought to (no doubt for further reasons), and he must believe that by nailing the boards together he will advance his project. Reference to other attitudes besides wanting, or thinking he ought, may help specify the agent's reasons, but it seems that some positive, or pro-, attitude must be involved. When we talk of reasons in this way, we do not require that the reasons be good ones. We learn something about a man's reasons for starting a war when we learn that he did it with the intention of ending all wars, even if we know that his belief that starting a war would end all wars was false. Similarly, a desire to humiliate an acquaintance may be someone's reason for cutting him at a party though an observer might, in a more normative vein, think that that was no reason. The falsity of a belief, or the patent wrongness of a value or desire, does not disqualify the belief or desire from providing an explanatory reason. On the other hand, beliefs and desires tell us an agent's reasons for acting only if those attitudes are appropriately related to the action as viewed by the actor. To serve as reasons for an action, beliefs and desires need not be reasonable, but a normative element nevertheless enters, since the action must be reasonable in the light of the beliefs and desires (naturally it may not be reasonable in the light of further considerations).

What does it mean to say that an action, as viewed by the agent, is reasonable in the light of his beliefs and desires? Suppose that a man boards an aeroplane marked 'London' with the intention of boarding an aeroplane headed for London, England. His reasons for boarding the plane marked 'London' are given by his desire to board a plane headed for London, England, and his belief that the plane marked 'London' is headed for London, England. His reasons explain why he intentionally boarded the plane marked 'London'. As it happens, the plane marked 'London' was headed for London, Ontario, not London, England, and so his reasons cannot explain why he boarded a plane headed for London, England. They can explain why he boarded a plane headed for London, Ontario, but only when the reasons are conjoined to the fact that the plane marked 'London' was headed for London, Ontario; and of course his reasons cannot explain why he intentionally boarded a plane headed for London, Ontario, since he had no such intention.[1]

The relation between reasons and intentions may be appreciated by comparing these statements:

(1) His reason for boarding the plane marked 'London' was that he wanted to board a plane headed for London, England, and he believed the plane marked 'London' was headed for London, England.

(2) His intention in boarding the plane marked 'London' was to board a plane headed for London, England.

---

[1] I take the 'intentionally' to govern the entire phrase 'boarded a plane headed for London, Ontario'. On an alternative reading, only the boarding would be intentional. Similarly, in (1) below his reason extends to the marking on the plane.

The first of these sentences entails the second, but not conversely. The failure of the converse is due to two differences between (1) and (2). First, from (2) it is not possible to reconstruct the specific pro attitude mentioned in (1). Given (2), there must be some appropriate pro attitude, but it does not have to be wanting. And second, the description of the action ('boarding the plane marked "London"') occupies an opaque context in (1), but a transparent context in (2). Thus 'boarding the plane headed for London, Ontario' describes the same action as 'boarding the plane marked "London"', since the plane marked 'London' *was* the plane headed for London, Ontario. But substitution of 'boarding the plane headed for London, Ontario' for 'boarding the plane marked "London"' will turn (1) false, while leaving (2) true. Of course the description of the intention in (2), like the description of the contents of the belief and pro attitude in (1), occupies an opaque context.

Finally, there is this relation between statements with the forms of (1) and (2): although (2) does not entail (1), if (2) is true, *some* statements with the form of (1) is true (with perhaps another description of the action, and with an appropriate pro attitude and belief filled in). Statement (1), unlike (2), must describe the agent's action in a way that makes clear a sense in which the action was reasonable in the light of the agent's reasons. So we can say, if an agent does *A* with the intention of doing *B*, there is some description of *A* which reveals the action as reasonable in the light of reasons the agent had in performing it.

When is an action (described in a particular way) reasonable in the light of specific beliefs and pro attitudes? One way to approach the matter is through a rather abstract account of practical reasoning. We cannot suppose that whenever an agent acts intentionally he goes through a process of deliberation or reasoning, marshals evidence and principles, and draws conclusions. Nevertheless, if someone acts with an intention, he must have attitudes and beliefs from which, had he been aware of them and had the time, he *could* have reasoned that his action was desirable (or had some other positive attribute). If we can characterize the reasoning that would serve, we will in effect have described the logical relations between descriptions of beliefs and desires, and the description of an action, when the former give the reasons with which the latter was performed. We are to imagine, then, that the agent's beliefs and desires provide him with the premises of an argument. In the case of belief, it is clear at once what the premise is. Take an example: someone adds sage to the stew with the intention of improving the taste. So his corresponding premise is: Adding sage to the stew will improve its taste.

The agent's pro attitude is perhaps a desire or want; let us suppose he wants to improve the taste of the stew. But what is the corresponding premise? If we were to look for the proposition toward which his desire is directed, the proposition he wants true, it would be something like: He does something that improves the taste of the stew (more briefly: He improves the taste of the stew). This cannot be his premise, however, for nothing interesting follows from the two premises: Adding sage to the stew will improve its taste, and the agent improves the taste of the stew. The trouble is that the attitude of *approval* which the agent has toward the second proposition has been left out. It cannot be put back in by making the premise 'The agent wants to improve the taste of the stew': we do not want a *description* of his desire, but an *expression* of it in a form in which he might use it to arrive at an action. The natural expression of his desire is, it seems to me, evaluative in form; for example, 'It is desirable to improve the taste of the stew,' or, 'I ought to improve the taste of the stew'. We may suppose different pro attitudes are expressed with other evaluative words in place of 'desirable'.

There is no short proof that evaluative sentences express desires and other pro attitudes in the way that the sentence 'Snow is white' expresses the belief that snow is white. But the following consideration will perhaps help show what is involved. If someone who knows English says honestly 'Snow is white', then he believes snow is white. If my thesis is correct, someone who says honestly 'It is desirable that I stop smoking' has some pro attitude towards his stopping smoking. He feels some inclination to do it; in fact he will do it if nothing stands in the way, he knows how, and he has no contrary values or desires. Given this assumption, it is reasonable to generalize: if explicit value judgements represent pro attitudes, all pro attitudes may be expressed by value judgements that are at least implicit.

This last stipulation allows us to give a uniform account of acting with an intention. If someone performs an action of type A with the intention of performing an action of type B, then he must have a pro attitude toward actions of type B (which may be expressed in the form: an action of type B is good (or has some other positive attribute)) and a belief that in performing an action of type A he will be (or probably will be) performing an action of type B (the belief may be expressed in the obvious way). The expressions of the belief and desire entail that actions of type A are, or probably will be, good (or desirable, just, dutiful, etc.). The description of the action provided by the phrase substituted for 'A' gives the description under which the desire and the belief rationalize the action. So to bring things back to our example, the desire to improve the taste of the stew and the belief that adding sage to the stew will improve its taste serve to rationalize an action described as 'adding sage to the stew'. (This more or less standard account of practical reasoning will be radically modified presently.)

There must be such rationalizing beliefs and desires if an action is done for a reason, but of course the presence of such beliefs and desires when the action is done does not suffice to ensure that what is done is done with the appropriate intention, or even with any intention at all. Someone might want tasty stew and believe sage would do the trick and put in sage thinking it was parsley; or put in sage because his hand was joggled. So we must add that the agent put in the sage because of his reasons. This 'because' is a source of trouble; it implies, so I believe, and have argued at length, the notion of cause. But not any causal relation will do, since an agent might have attitudes and beliefs that would rationalize an action, and they might cause him to perform it, and yet because of some anomaly in the causal chain, the action would not be intentional in the expected sense, or perhaps in any sense.[2]

We end up, then, with this incomplete and unsatisfactory account of acting with an intention: an action is performed with a certain intention if it is caused in the right way by attitudes and beliefs that rationalize it.[3]

If this account is correct, then acting with an intention does not require that there be any mysterious act of the will or special attitude or episode of willing. For the account needs only desires (or other pro attitudes), beliefs, and the actions themselves. There is indeed the relation between these, causal or otherwise, to be anal-

---

[2] See Essay 4.

[3] This is where Essay 1 left things. At the time I wrote it I believed it would be possible to characterize 'the right way' in non-circular terms.

ysed, but that is not an embarrassing entity that has to be added to the world's furniture. We would not, it is true, have shown how to *define* the concept of acting with an intention; the reduction is not definitional but ontological. But the ontological reduction, if it succeeds, is enough to answer many puzzles about the relation between the mind and the body, and to explain the possibility of autonomous action in a world of causality.

This brings me back to the problem I mentioned at the start, for the strategy that appears to work for acting with an intention has no obvious application to pure intending, that is, intending that is not necessarily accompanied by action. If someone digs a pit with the intention of trapping a tiger, it is perhaps plausible that no entity at all, act, event, or disposition, corresponds to the noun phrase, 'the intention of trapping a tiger'—this is what our survey has led us to hope. But it is not likely that if a man has the intention of trapping a tiger, his intention is not a state, disposition, or attitude of some sort. Yet if this is so, it is quite incredible that this state or attitude (and the connected event or act of *forming an intention*) should play no role in acting with an intention. Our inability to give a satisfactory account of pure intending on the basis of our account of intentional action thus reflects back on the account of intentional action itself. And I believe the account I have outlined will be seen to be incomplete when we have an adequate analysis of pure intending.

Of course, we perform many intentional actions without forming an intention to perform them, and often intentional action is not preceded by an intention. So it would not be surprising if something were present in pure intending that is not always present in intentional action. But it would be astonishing if that extra element were foreign to our understanding of intentional action. For consider some simple action, like writing the word 'action'. Some temporal segments of this action are themselves actions: for example, first I write the letter 'a'. This I do with the intention of initiating an action that will not be complete until I have written the rest of the word. It is hard to see how the attitude towards the complete act which I have as I write the letter 'a' differs from the pure intention I may have had a moment before. To be sure, my intention has now begun to be realized, but why should that necessarily change my attitude? It seems that in any intentional action that takes much time, or involves preparatory steps, something like pure intending must be present.

We began with pure intending—intending without conscious deliberation or overt consequence—because it left no room for doubt that intending is a state or event separate from the intended action or the reasons that prompted the action. Once the existence of pure intending is recognized, there is no reason not to allow that intention of exactly the same kind is also present when the intended action eventuates. So though I may, in what follows, seem sometimes to concentrate on the rather special case of unfulfilled intentions, the subject is, in fact, all intending—intending abstracted from a context which may include any degree of deliberation and any degree of success in execution. Pure intending merely shows that there is something there to be abstracted.

What success we had in coping with the concept of intentional action came from treating talk of the intention with which an action is done as talk of beliefs, desires, and action. This suggests that we try treating pure intentions—intendings abstracted from normal outcomes—as actions, beliefs or pro-attitudes of some sort. The rest of this Essay is concerned with these possibilities.

Is pure intending an action? It may be objected that intending to do something is not a change or event of any kind, and so cannot be something the agent does. But this objection is met by an adjustment in the thesis; we should say that the action is forming an intention, while pure intending is the state of an agent who has formed an intention (and has not changed his mind). Thus all the weight is put on the idea of forming an intention. It will be said that most intentions are not formed, at least if forming an intention requires conscious deliberation or decision. What we need then is the broader and more neutral concept of coming to have an intention—a change that may take place so slowly or unnoticed that the agent cannot say when it happens. Still, it is an event, and we could decide to call it an action, or at least something the agent does.

I see no reason to reject this proposal; the worst that can be said of it is that it provides so little illumination. The state of intention just is what results from coming to have an intention—but what sort of a state is it? The coming to have an intention we might try connecting with desires and beliefs as we did other intentional actions (again with a causal chain that works 'in the right way'). But the story does not have the substantial quality of the account of intentional action because the purported action is not familiar or observable, even to the agent himself.

Another approach focuses on overt speech acts. *Saying* that one intends to do something, or that one will do it, is undeniably an action and it has some of the characteristics of forming an intention. Saying, under appropriate circumstances, that one intends to do something, or that one will do it, can commit one to doing it; if the deed does not follow, it is appropriate to ask for an explanation. Actually to identify saying one intends to do something with forming an intention would be to endorse a sort of performative theory of intention; just as saying one promises to do something may be promising to do it, saying one intends to do it may be intending (or forming the intention) to do it. Of course one may form an intention without saying anything aloud, but this gap may be filled with the notion of speaking to oneself, 'saying in one's heart'.[4] A variant theory would make forming an intention like (or identical with) addressing a command to oneself.

I think it is easy to see that forming an intention is quite different from saying something, even to oneself, and that intending to do something is quite different from having said something. For one thing, the performative character of commands and promises which makes certain speech acts surprisingly momentous depends on highly specific conventions, and there are no such conventions governing the formation of intentions. Promising involves assuming an obligation, but even if there are obligations to oneself, intending does not normally create one. If an agent does not do what he intended to do, he does not normally owe himself an explanation or apology, especially if he simply changed his mind; yet this is just the case that calls for explanation or apology when a promise has been made to another and broken. A command may be disobeyed, but only while it is in force. But if an agent does not do what he intended because he has changed his mind, the original intention is no longer in force. Perhaps it is enough to discredit these theories to point out that promising and commanding, as we usually understand them, are necessarily public performances, while forming an intention is not. Forming an intention may be an action but it is not a performance, and having an intention is not generally the aftermath of one.

[4] See P. T. Geach, *Mental Acts*.

None of this is to deny that saying 'I intend to do it' or 'I will do it' is much like, or on occasion identical with, promising to do it. If I say any of these things in the right context, I entitle a hearer to believe I believe I will do it. Perhaps a simpler way to put it is this: if I say 'I intend to do it' or 'I will do it' or 'I promise to do it' under certain conditions, then I *represent myself* as believing that I will. I may not believe I will, I may not intend that my hearer believe I will, but I have given him ground for complaint if I do not. These facts suggest that if I not only say 'I intend' or 'I will' in such a way as to represent myself as believing I will, but I am sincere as well, then my sincerity guarantees both that I intend to do it and that I believe I will. Some such line of argument has led many philosophers to hold that intending to do something entails believing one will, and has led a few philosophers to the more extreme view that to intend to do something is identical with a sort of belief that one will.

Is intending to act a belief that one will? The argument just sketched does not even show that intending implies belief. The argument proves that a man who sincerely says 'I intend to do it' or 'I will do it' under certain conditions must believe he will do it. But it may be the saying, not the intention, that implies the belief. And I think we can see this is the case. The trouble is that we have asked the notion of sincerity to do two different pieces of work. We began by considering cases where, by saying 'I intend to' or 'I will', I entitle a hearer to believe I will. And here it is obvious that if I am sincere, if things are as I represent them, then I must believe I will. But it is an assumption unsupported by the argument that any time I sincerely say I intend to do something I must believe I will do it, for sincerity in this case merely requires that I know I intend to do it. We are agreed that there are cases where sincerity in the utterer of 'I intend to' requires him to believe he will, but the argument requires that these cases include all those in which the speaker knows or believes he intends to do it.

Once we have distinguished the question how belief is involved in avowals of intention from the question how belief is involved in intention, we ought to be struck with how dubious the latter connection is.

It is a mistake to suppose that if an agent is doing something intentionally, he must know that he is doing it. For suppose a man is writing his will with the intention of providing for the welfare of his children. He may be in doubt about his success and remain so to his death; yet in writing his will he may in fact be providing for the welfare of his children, and if so he is certainly doing it intentionally. Some sceptics may think this example fails because they refuse to allow that a man may *now* be providing for the welfare of his children if that welfare includes events yet to happen. So here is another example: in writing heavily on this page I may be intending to produce ten legible carbon copies. I do not know, or believe with any confidence, that I am succeeding. But if I am producing ten legible carbon copies, I am certainly doing it intentionally. These examples do not prove that pure intending may not imply belief, for the examples involve acting with an intention. Nevertheless, it is hard to imagine that the point does not carry over to pure intending. As he writes his will, the man not only is acting with the intention of securing the welfare of his children, but he also intends to secure the welfare of his children. If he can be in doubt whether he is now doing what he intends, surely he can be in doubt whether he will do what he intends.

The thesis that intending implies believing is sometimes defended by claiming that expressions of intention are generally incomplete or elliptical. Thus the man writing his will should be described as intending to try to secure the welfare of his children, not as intending to secure it, and the man with the carbon paper is merely intending to try to produce his copies. The phrases sound wrong: we should be much more apt to say he *is* trying, and intends to do it. But where the action is entirely in the future, we do sometimes allow that we intend to try, and we see this as more accurate than the bald statement of intention when the outcome is sufficiently in doubt. Nevertheless, I do not think the claim of ellipsis can be used to defend the general thesis.

Without doubt many intentions are conditional in form—one intends to do something only if certain conditions are satisfied—and without doubt we often suppress mention of the conditions for one reason or another. So elliptical statements of intention are common. Grice gives us this exchange:

X. I intend to go to that concert on Tuesday.
Y. You will enjoy that.
X. I may not be there.
Y. I am afraid I don't understand.
X. The police are going to ask me some awkward questions on Tuesday afternoon, and I may be in prison by Tuesday evening.

Y. Then you should have said to begin with, 'I intend to go to the concert if I am not in prison', or, if you wished to be more reticent, something like, 'I should probably be going', or 'I hope to go', or, 'I aim to go', or, 'I intend to go if I can'.[5]

Grice does not speak of ellipsis here but he does think that this example, and others like it, make a strong case for the view that 'X intends to do *A*' is true, when 'intends' is used in the *strict* sense, only if X is sure that he will do *A*. The man in the example must intend *something*, and so if we knew what it was, we could say that his remark 'I intend to go to the concert' was elliptical for what he would have said if he had used 'intend' in the strict sense. What would he have said? 'I hope to go' is not more accurate about the intention, since it declares no intention at all; similarly for 'I aim to go' and 'I should probably be going', 'I intend to go if I can' is vague and general given the particularity of X's doubts, but there seems something worse wrong with it. For if an agent cannot intend what he believes to be impossible, then he asserts neither more nor less by saying 'I intend to do it if I can' than he would by saying 'I intend to do it.' How about 'I intend to go to the concert if I am not in prison'? Intuitively, this comes closest to conveying the truth about the situation as X sees it. But is it *literally* more accurate? It is hard to see how. On the view Grice is arguing for, if X said in the strict sense, and honestly, 'I intend to be at the concert', he would imply that he believed he would be there. If X said in the strict sense, and honestly, 'I intend to be at the concert if I am not in jail', he would imply that he believed he would be at the concert if he were not in jail. Now obviously the first belief implies the second but is not implied by it, and so an expression of the second belief makes a lesser claim and may be thought to be more accurate. Of course, the stronger claim cannot, by its contents, lead Y into error about what X will do, for whether X says he will be at the concert, or only that he will be there if he is not in jail, both X and Y know X will not be at the concert if he is in jail. Where Y might be misled, if the thesis we are examining is true, is with respect to what X believes he will do, and hence what he intends. For on the thesis, 'I intend to be at the concert if I am not in jail' implies a weaker belief than 'I intend to be at the concert'. If this is right, then greater accuracy still would result from further provisos, since X also does not believe he will be at the

[5] H. P. Grice, 'Intention and Uncertainty', 4–5.

concert if he changes his mind, or if something besides imprison-ment prevents him. We are thus led further and further toward the nearly empty, 'I intend to do it if nothing prevents me, if I don't change my mind, if nothing untoward happens'. This tells us almost nothing about what the agent believes about the future, or what he will in fact do.

I think X spoke correctly and accurately, but misleadingly, when he said, 'I intend to go to the concert'. He could have corrected the impression while still being accurate by saying, 'I now intend to go to the concert, but since I may be put in jail, I may not be there'. A man who says, 'I intend to be there, but I may not be' does not contradict himself, he is at worst inscrutable until he says more. We should realize there is something wrong with the idea that most statements of intention are elliptical until tempered by our doubts about what we shall in fact do when we notice that there is no satisfactory *general* method for supplying the more accurate state-ment for which the original statement went proxy. And the reason is clear: there can be no finite list of things we think might prevent us from doing what we intend, or of circumstances that might cause us to stay our hand. If we are reasonably sure something will prevent us from acting, this does, perhaps, baffle intention, but if we are simply uncertain, as is often the case, intention is not necessarily dulled. We can be clear what it is we intend to do while being in the dark as to the details, and therefore the pitfalls. If this is so, being more accurate about what we intend cannot be a matter of being more accurate about what we believe we will bring off.

There are genuine conditional intentions, but I do not think they come in the form, 'I intend to do it if I can' or, 'if I don't change my mind'. Genuine conditional intentions are appropriate when we explicitly consider what to do in various contingencies; for example, someone may intend to go home early from a party if the music is too loud. If we ask for the difference between conditions that really do make the statement of an intention more accurate and bogus conditions like 'if I can' or 'if nothing comes up' or 'if I don't change my mind', it seems to me clear that the difference is this: bona fide conditions are ones that are reasons for acting that are contem-porary with the intention. Someone may not like loud music now, and that may be why he now intends to go home early from the party if the music is too loud. His not being able to go home early is not a reason for or against his going home early, and so it is not a relevant condition for an intention, though if he believes he cannot do it, that may prevent his having the intention. Changing his mind is a tricky case, but in general someone is not apt to view a possible future change of intention as a reason to modify his present intention unless he thinks the future change will itself be brought about by something he would now consider a reason.

The contrast that has emerged, between the circumstances we do sometimes allow to condition our intentions and the circumstances we would allow if intentions implied the belief that we will do what we intend, seems to me to indicate pretty conclusively that we do not necessarily believe we will do what we intend to do, and that we do not state our intentions more accurately by making them conditional on all the circumstances in whose presence we think we would act.

These last considerations point to the strongest argument against identifying pure intending with belief one will do what one intends. This is that reasons for intending to do something are in general quite different from reasons for believing one will do it. Here is why I intend to reef the main: I see a squall coming, I want to prevent the boat from capsizing, and I believe that reefing the main will prevent the boat from capsizing. I would put my reasons for intending to reef the main this way: a squall is coming, it would be a shame to capsize the boat, and reefing the main will prevent the boat from capsizing. But these reasons for intending to reef the main in themselves give me no reason to believe I will reef the main. Given additional assumptions, for example, that the approach of a squall is a reason to believe I believe a squall is coming, and that the shamefulness of capsizing the boat may be a reason to believe I want to prevent the boat from capsizing; and given that I have these beliefs and desires, it may be reasonable to suppose I intend to reef the main, and will in fact do so. So there may be a loose connection between reasons of the two kinds, but they are not at all identical (individual reasons may be the same, but a smallest natural set of reasons that supports the intention to act cannot be a set that supports the belief that the act will take place).

It is often maintained that an intention is a belief not arrived at by reasoning from evidence, or that an intention is a belief about one's future action that differs in some other way in its origin from an ordinary prediction. But such claims do not help the thesis. How someone arrived at a belief, what reasons he would give in support of it, what sustains his faith, these are matters that are simply irrelevant to the question what constitute reasons for the belief; the former events are accidents that befall a belief and cannot change its logical status without making it a new belief.

Is intending to do something the same as wanting to do it? Clearly reasons for intending to do something are very much like reasons for action, indeed one might hold that they are exactly the same except for time. As John Donne says, 'To will implies delay', but we may reduce the delay to a moment. I am writing the letter 'a' of 'action', and I intend to write the letter 'c' as soon as I finish the 'a'. The reason I intend to write the letter 'c' as soon as I finish the 'a' is that I want to write the word 'action', and I know that to do this I must, after writing the letter 'a', write the letter 'c'. Now I have finished the 'a' and have begun 'c'! What is my reason for writing the 'c'? It is that I want to write the word 'action', and I know that to this I must ... So far the reasons sound identical, but if we look closer a difference will emerge. When I am writing the 'a' I intend to write the 'c' in just a moment, and part of my reason is that I believe this moment looms in the immediate future; when I am writing the 'c', my reasons include the belief that *now* is the time to write the 'c' if I am to write 'action', as I wish to. Aristotle sometimes neglects this difference and as a result says things that sound fatuous. He is apt to give as an example of practical reasoning something of this sort: I want to be warm, I believe a house will keep me warm, straightway I build a house. It is an important doctrine that the conclusion of a piece of practical reasoning may be an action; it is also important that the conclusion may be the formation of an intention to do something in the future.

Now I would like to draw attention to an aspect of this picture of what it is like to form an intention that seems to make for a difficulty. Consider again a case of intentional action. I want to eat something sweet, that is, I hold that my eating something sweet is desirable. I believe this candy is sweet, and so my eating this candy will be a case of my eating something sweet, and I conclude that my eating this candy is desirable. Since nothing stands in the way, I eat the candy—the conclusion is the action. But this also means I could express the conclusion by using a demonstrative reference to the action: 'This action of mine, this eating by me of this candy now, is desirable'. What seems so important about the possibility of a demonstrative reference to the action is that it is a case where it makes sense to couple a value judgement directly to action. My evaluative reason for acting was, 'My eating something sweet is desirable'. But of course this cannot mean that any action of mine whatsoever that is an eating of something sweet is something it makes sense to do—my judgement merely deals with actions in so far as they are sweet-consuming. Some such actions, even all of them, may have plenty else wrong with them. It is only when I come to an actual action that it makes sense to judge it as a whole as desirable or not; up until that moment there was no object with which I was acquainted to judge. Of course I can still say of the completed action that it is desirable in so far as it is this or that, but in choosing to perform it I went beyond this; my choice represented, or perhaps was, a judgement that the action itself was desirable.

And now the trouble about pure intending is that there is no action to judge simply good or desirable. All we can judge at the stage of pure intending is the desirability of actions of a sort, and actions of a sort are generally judged on the basis of the aspect that defines the sort. Such judgements, however, do not always lead to reasonable action, or we would be eating everything sweet we could lay our hands on (as Anscombe pointed out).[6]

The major step in clearing up these matters is to make a firm distinction between the kind of judgement that corresponds to a desire like wanting to eat something sweet and the kind of judgement that can be the conclusion of a piece of practical reasoning—that can correspond to an intentional action.[7] The first sort of judgement is often thought to have the form of a law: any action that is an eating of something sweet is desirable. If practical reasoning is deductive, this is what we should expect (and it seems to be how Aristotle and Hume, for example, thought of practical reasoning). But there is a fundamental objection to this idea, as can be seen when we consider an action that has both a desirable and an undesirable aspect. For suppose the propositional expression of a desire to eat something sweet were a universally quantified conditional.

[6] G. E. M. Anscombe, *Intention*, 59.
[7] No weight should be given the word 'judgement'. I am considering here the *form* of propositions that express desires and other attitudes. I do not suppose that someone who wants to eat something sweet necessarily *judges* that it would be good to eat something sweet; perhaps we can say he *holds* that his eating something sweet has some positive characteristic. By distinguishing among the propositional expressions of attitudes I hope to mark differences among the attitudes.

While holding it desirable to eat something sweet, we may also hold that it is undesirable to eat something poisonous. But one and the same object may be sweet and poisonous, and so one and the same action may be the eating of something sweet and of something poisonous. Our evaluative principles, which seem consistent, can then lead us to conclude that the same action is both desirable and undesirable. If undesirable actions are not desirable, we have derived a contradiction from premises all of which are plausible. The cure is to recognize that we have assigned the wrong form to evaluative principles. If they are judgements to the effect that *in so far as* an action has a certain characteristic it is good (or desirable, etc.), then they must not be construed in such a way that detachment works, or we will find ourselves concluding that the action is simply desirable when all that is warranted is the conclusion that it is desirable in a certain respect. Let us call judgements that actions are desirable in so far as they have a certain attribute, prima facie judgements.

Prima facie judgements cannot be directly associated with actions, for it is not reasonable to perform an action merely because it has a desirable characteristic. It is a reason for acting that the action is believed to have some desirable characteristic, but the fact that the action is performed represents a further judgement that the desirable characteristic was enough to act on—that other considerations did not outweigh it. The judgement that corresponds to, or perhaps is identical with, the action cannot, therefore, be a prima facie judgement; it must be an all-out or unconditional judgement which, if we were to express it in words, would have a form like 'This action is desirable'.

It can now be seen that our earlier account of acting with an intention was misleading or at least incomplete in an important respect. The reasons that determine the description under which an action is intended do not allow us to *deduce* that the action is simply worth performing; all we can deduce is that the action has a feature that argues in its favour. This is enough, however, to allow us to give the intention with which the action was performed. What is misleading is that the reasons that enter this account do not generally constitute all the reasons the agent considered in acting, and so knowing the intention with which someone acted does not allow us to reconstruct his actual reasoning. For we may not know how the agent got from his desires and other attitudes—his prima facie reasons—to the conclusion that a certain action was desirable.[8]

In the case of intentional action, at least when the action is of brief duration, nothing seems to stand in the way of an Aristotelian identification of the action with a judgement of a certain kind—an all-out, unconditional judgement that the action is desirable (or has some other positive characteristic). The identification of the action with the conclusion of a piece of practical reasoning is not essential to the view I am endorsing, but the fact that it can be made explains why, in our original account of intentional action, what was needed to relate it to pure intending remained hidden.

In the case of pure intending, I now suggest that the intention simply is an all-out judgement. Forming an intention, deciding, choosing, and deliberating are various modes of arriving at the judgement, but it is possible to come to have such a judgement or attitude without any of these modes applying.

Let me elaborate on this suggestion and try to defend it against some objections. A few pages ago I remarked that an all-out judgement makes sense only when there is an action present (or past) that is known by acquaintance. Otherwise (I argued) the judgement must be general, that is, it must cover all actions of a certain sort, and among these there are bound to be actions some of which are desirable and some not. Yet an intention cannot single out a particular action in an intelligible sense, since it is directed to the future. The puzzle arises, I think, because we have overlooked an important distinction. It would be mad to hold that any action of mine in the immediate future that is the eating of something sweet would be desirable. But there is nothing absurd in my judging that any action of mine in the immediate future that is the eating of something sweet would be desirable *given the rest of what I believe about the immediate future*. I do not believe I will eat a poisonous candy, and so that is not one of the actions of eating something sweet that my all-out judgement includes. It would be a mistake to try to improve the statement of my intention by saying, 'I intend to eat something sweet, provided it isn't poisonous'. As we saw, this is a mistake because if this is the road I must travel, I will never get my intentions right. There are *endless* circumstances under which I would not eat something sweet, and I cannot begin to foresee them all. The point

[8] I have said more about the form of prima facie evaluative judgements, and the importance of distinguishing them from unconditional judgements, in Essay 2.

is, I do not believe anything will come up to make my eating undesirable or impossible. That belief is not part of what I intend, but an assumption without which I would not have the intention. The intention is not conditional in form; rather, the existence of the intention is conditioned by my beliefs.

I intend to eat a hearty breakfast tomorrow. You know, and I know, that I will not eat a hearty breakfast tomorrow if I am not hungry. And I am not certain I will be hungry, I just think I will be. Under these conditions it is not only not more accurate to say, 'I intend to eat a hearty breakfast if I'm hungry', it is *less* accurate. I have the second intention as well as the first, but the first implies the second, and not vice versa, and so the first is a more complete account of my intentions. If you knew only that I intended to eat a hearty breakfast if I was hungry, you would not know that I believe I will be hungry, which is actually the case. But you might figure this out if you knew I intend to eat a hearty breakfast tomorrow.

I think this view of the matter explains the trouble we had about the relation between intending to do something and believing one will—why, on the one hand, it is so strange to say, 'I intend to do it but perhaps I won't', and yet, on the other hand, it is so impossible to increase the accuracy of statements of intention by making the content of the intention conditional on how things turn out. The explanation is that the intention assumes, but does not contain a reference to, a certain view of the future. A present intention with respect to the future is in itself like an interim report: given what I now know and believe, here is my judgement of what kind of action is desirable. Since the intention is based on one's best estimate of the situation, it merely distorts matters to say the agent intends to act in the way he does only if his estimate turns out to be right. A present intention does not need to be anything like a resolve or a commitment (though *saying* one intends to do something may sometimes have this character). My intention is based on my present view of the situation; there is no reason in general why I should act as I now intend if my present view turns out to be wrong.

We can now see why adding, 'if I can' never makes the statement of an intention more accurate, although it may serve to cancel an unwanted natural suggestion of the act of saying one intends to do something. To intend to perform an action is, on my account, to hold that it is desirable to perform an action of a certain sort in the light of what one believes is and will be the case. But if one believes no such action is possible, then there can be no judgement that such an action consistent with one's beliefs is desirable. There can be no such intention.

If an intention is just a judgement that an action of a certain sort is desirable, what is there to distinguish an intention from a mere wish? We may put aside wishes for things that are not consistent with what one believes, for these are ruled out by our conception of an intention. And we may put aside wishes that do not correspond to all-out judgements. ('I wish I could go to London next week': my going to London next week may be consistent with all I believe, but not with all I want. This wish is idle because it is based on some only of my prima facie reasons.) But once we put these cases aside, there is no need to distinguish intentions from wishes. For a judgement that something I think I can do, and that I think I see my way clear to doing, a judgement that such an action is desirable not only for one or another reason but in the light of all my reasons; a judgement like this is not a mere wish. It *is* an intention. (This is not to deny that there are borderline cases.)

How well have we coped with the problem with which we began? That problem was, in effect, to give an account of intending (and of forming an intention) that would mesh in a satisfactory way with our account of acting with an intention and would not sacrifice the merits of that account. With respect to the first point, finding an account of intending that would mesh with our account of intentional action, we devised a satisfactory way of relating the two concepts, but only by introducing a new element, an all-out judgement, into the analysis of intentional action. Given this sort of judgement and the idea of such a judgement made in the light of what is believed about the future course of affairs, we were able, I think, to arrive at a plausible view of intending.

There remains the question whether the sort of judgement to which I have referred, an all-out judgement, can be understood without appeal to the notions of intention or will. I asked at the beginning of this last section of my paper whether intending to do something is wanting to do it; if it were, we might consider that our aim had been achieved. What we intend to do we want, in some very broad sense of want, to do. But this does not mean that intending is a form of wanting. For consider the actions that I want to perform and that are consistent with what I believe. Among them are the actions I intend to perform, and many more. I want to go to London next week, but I do not intend to, not because I think I cannot, but because it would interfere with other things I want more. This suggests strongly that wanting and desiring are best viewed as corresponding to, or constituting, prima facie judgements.

If this is correct, we cannot claim that we have made out a case for viewing intentions as something familiar, a kind of wanting, where we can distinguish the kind without having to use the concept of intention or will. What we can say, however, is that intending and wanting belong to the same genus of pro attitudes expressed by value judgements. Wants, desires, principles, prejudices, felt duties, and obligations provide reasons for actions and intentions, and are expressed by prima facie judgements; intentions and the judgements that go with intentional actions are distinguished by their all-out or unconditional form. Pure intendings constitute a subclass of the all-out judgements, those directed to future actions of the agent, and made in the light of his beliefs.

EVENT AND CAUSE

6 *The Logical Form of Action Sentences*

Strange goings on! Jones did it slowly, deliberately, in the bathroom, with a knife, at midnight. What he did was butter a piece of toast. We are too familiar with the language of action to notice at first an anomaly: the 'it' of 'Jones did it slowly, deliberately, . . .' seems to refer to some entity, presumably an action, that is then characterized in a number of ways. Asked for the logical form of this sentence, we might volunteer something like, 'There is an action x such that Jones did x slowly and Jones did x deliberately and Jones did x in the bathroom, . . .' and so on. But then we need an appropriate singular term to substitute for 'x'. In fact we know Jones buttered a piece of toast. And, allowing a little slack, we can substitute for 'x' and get 'Jones buttered a piece of toast slowly and Jones buttered a piece of toast deliberately and Jones buttered a piece of toast in the bathroom . . .' and so on. The trouble is that we have nothing here we would ordinarily recognize as a singular term. Another sign that we have not caught the logical form of the sentence is that in this last version there is no implication that any *one* action was slow, deliberate, and in the bathroom, though this is clearly part of what is meant by the original.

   The present Essay is devoted to trying to get the logical form of simple sentences about actions straight. I would like to give an account of the logical or grammatical role of the parts or words of such sentences that is consistent with the entailment relations between such sentences and with what is known of the role of those same parts or words in other (non-action) sentences. I take this enterprise to be the same as showing how the meanings of action sentences depend on their structure. I am not concerned with the meaning analysis of logically simple expressions in so far as this goes beyond the question of logical form. Applied to the case at hand, for example, I am not concerned with the meaning of 'deliberately' as opposed, perhaps, to 'voluntary'; but I am interested in the logical role of both these words. To give another illustration of the distinction I have in mind: we need not view the difference between 'Joe believes that there is life on Mars' and 'Joe knows that there is life on Mars' as a difference in logical form. That the second, but not the first, entails 'There is life on Mars' is plausibly a logical truth; but it is a truth that emerges only when we consider the meaning analysis of 'believes' and 'knows'. Admittedly there is something arbitrary in how much of logic to pin on logical form. But limits are set if our interest is in giving a coherent and constructive account of meaning: we must uncover enough structure to make it possible to state, for an arbitrary sentence, how its meaning depends on that structure, and we must not attribute more structure than such a theory of meaning can accommodate.

Consider the sentence:

(1) Jones buttered the toast slowly, deliberately, in the bathroom, with a knife, at midnight.

Despite the superficial grammar we cannot, I shall argue later, treat the 'deliberately' on a par with the other modifying clauses. It alone imputes intention, for of course Jones may have buttered the toast slowly, in the bathroom, with a knife, at midnight, and quite unintentionally, having mistaken the toast for his hairbrush which was what he intended to butter. Let us, therefore, postpone discussion of the 'deliberately' and its intentional kindred.

'Slowly', unlike the other adverbial clauses, fails to introduce a new entity (a place, an instrument, a time), and also may involve a special difficulty. For suppose we take 'Jones buttered the toast slowly' as saying that Jones's buttering of the toast was slow; is it clear that we can equally well say of Jones's action, no matter how we describe it, that it was slow? A change in the example will help. Susan says, 'I crossed the Channel in fifteen hours.' 'Good grief, that was slow.' (Notice how much more naturally we say 'slow' here than 'slowly'. But *what* was slow, what does 'that' refer to? No appropriate singular term appears in 'I crossed the Channel in fifteen hours.') Now Susan adds, 'But I swam.' 'Good grief, that was fast.' We do not withdraw the claim that it was a slow crossing; this is consistent with its being a fast swimming. Here we have enough to show, I think, that we cannot construe 'It was a slow crossing' as 'It was slow and it was a crossing' since the crossing may also be a swimming that was not slow, in which case we would have 'It was slow and it was a crossing and it was a swimming and it was not slow.' The problem is not peculiar to talk of actions, however. It appears equally when we try to explain the logical role of the attributive adjectives in 'Grundy was a short basketball player, but a tall man', and 'This is a good memento of the murder, but a poor steak knife.' The problem of attributives is indeed a problem about logical form, but it may be put to one side here because it is not a problem for action sentences alone.

We have decided to ignore, for the moment at least, the first two adverbial modifiers in (1), and may now deal with the problem of the logical form of:

(2) Jones buttered the toast in the bathroom with a knife at midnight.

Anthony Kenny, who deserves the credit for calling explicit attention to this problem,[1] points out that most philosophers today would, as a start, analyse this sentence as containing a five-place predicate with the argument places filled in the obvious ways with singular terms or bound variables. If we go on to analyse 'Jones buttered the toast' as containing a two-place predicate, 'Jones buttered the toast in the bathroom' as containing a three-place predicate, and so forth, we obliterate the logical relations between these sentences, namely that (2) entails the others. Or, to put the objection another way, the original sentences contain a common syntactic element ('buttered') which we intuitively recognize as relevant to the meaning relations of the sentences. But the proposed analyses show no such common element.

Kenny rejects the suggestion that 'Jones buttered the toast' be considered as elliptical for 'Jones buttered the toast somewhere with something at some time', which would restore the wanted entailments, on the ground that we could never be sure how many standby positions to provide in each predicate of action. For example, couldn't we add to (2) the phrase 'by holding it between the toes of his left foot'? Still, this adds a place to the predicate only if it differs in meaning from, 'while holding it between the toes of his

---

[1] Anthony Kenny, *Action, Emotion and Will*, Ch. VII.

left foot', and it is not quite clear that this is so. I am inclined to agree with Kenny that we cannot view verbs of action as usually containing a large number of standby positions, but I do not have what I consider a knock-down argument. (A knock-down argument would consist in a method for increasing the number of places indefinitely.)[2]

Kenny proposes that we may exhibit the logical form of (2) in somewhat the following manner:

(3) Jones brought it about that the toast was buttered in the bathroom with a knife at midnight.

Whatever the other merits in this proposal (I shall consider some of them presently) it is clear that it does not solve the problem Kenny raises. For it is, if anything, even more obscure how (3) entails 'Jones brought it about that the toast was buttered' or 'The toast was buttered' then how (2) entails 'Jones buttered the toast.' Kenny seems to have confused two different problems. One is the problem of how to represent the idea of *agency*: it is this that prompts Kenny to assign 'Jones' a logically distinguished role in (3). The other is the problem of the 'variable polyadicity' (as Kenny calls it) of action ver`s. And it is clear that this problem is independent of the first, since it arises with respect to the sentences that replace '*p*' in '*x* brings it about that *p*'.

If I say I bought a house downtown that has four bedrooms, two fireplaces, and a glass chandelier in the kitchen, it's obvious that I can go on forever adding details. Yet the logical form of the sentences I use presents no problem (in this respect). It is something like, 'There is a house such that I bought it, it is downtown, it has four bedrooms, . . .' and so forth. We can tack on a new clause at will because the iterated relative pronoun will carry the reference back to the same entity as often as desired. (Of course we know how to state this much more precisely.) Much of our talk of action suggests the same idea: that there are such *things* as actions, and that a sentence like (2) describes the action in a number of ways. 'Jones did it with a knife.' 'Please tell me more about it.' The 'it' here

---

[2] Kenny seems to think there is such a method, for he writes, 'If we cast our net widely enough, we can make "Brutus killed Caesar" into a sentence which describes, with a certain lack of specification, the whole history of the world, (op. cit., 160). But he does not show how to make each addition to the sentence one that irreducibly modifies the killing as opposed, say, to Brutus or Caesar, or the place or the time.

doesn't refer to Jones or the knife, but to what Jones did—or so it seems.

'. . . it is in principle always open to us, along various lines, to describe or refer to "what I did" in so many ways,' writes Austin.[3] Austin is obviously leery of the apparent singular term, which he puts in scare quotes; yet the grammar of his sentence requires a singular term. Austin would have had little sympathy, I imagine, for the investigation into logical form I am undertaking here, though the demand that underlies it, for an intuitively acceptable and constructive theory of meaning, is one that begins to appear in the closing chapters of *How to Do Things with Words*. But in any case, Austin's discussion of excuses illustrates over and over the fact that our common talk and reasoning about actions is most naturally analysed by supposing that there are such entities.

'I didn't know it was loaded' belongs to one standard pattern of excuse. I do not deny that I pointed the gun and pulled the trigger, nor that I shot the victim. My ignorance explains how it happened that I pointed the gun and pulled the trigger intentionally, but did not shoot the victim intentionally. That the bullet pierced the victim was a consequence of my pointing the gun and pulling the trigger. It is clear that these are two different events, since one began slightly after the other. But what is the relation between my pointing the gun and pulling the trigger, and my shooting the victim? The natural and, I think, correct answer is that the relation is that of identity. The logic of this sort of excuse includes, it seems, at least this much structure: I am accused of doing b, which is deplorable. I admit I did a, which is excusable. My excuse for doing b rests upon my claim that I did not know that $a = b$.

Another pattern of excuse would have me allow that I shot the victim intentionally, but in self-defence. Now the structure includes something more. I am still accused of b (my shooting the victim), which is deplorable. I admit I did c (my shooting the victim in self-defence), which is excusable. My excuse for doing b rests upon my claim that I knew or believed that $b = c$.

The story can be given another twist. Again I shoot the victim, again intentionally. What I am asked to explain is my shooting of the bank president (d), for the victim was that distinguished gentleman. My excuse is that I shot the escaping murderer (e), and surprising

---

[3] J. L. Austin, 'A Plea for Excuses', 148.

and unpleasant as it is, my shooting the escaping murderer and my shooting of the bank president were one and the same action $(e = d)$, since the bank president and the escaping murderer were one and the same person. To justify the 'since' we must presumably think of 'my shooting of x' as a functional expression that names an action when the 'x' is replaced by an appropriate singular term. The relevant reasoning would then be an application of the principle $x = y \rightarrow fx = fy$.

Excuses provide endless examples of cases where we seem compelled to take talk of 'alternative descriptions of the same action' seriously, i.e., literally. But there are plenty of other contexts in which the same need presses. *Explaining* an action by giving an intention with which it was done provides new descriptions of the action: I am writing my name on a piece of paper with the intention of writing a cheque with the intention of paying my gambling debt. List all the different descriptions of my action. Here are a few for a start: I am writing my name. I am writing my name on a piece of paper. I am writing my name on a piece of paper with the intention of writing a cheque. I am writing a cheque. I am paying my gambling debt. It is hard to imagine how we can have a coherent theory of action unless we are allowed to say that each of these sentences is made true by the same action. Redescription may supply the motive ('I was getting my revenge'), place the action in the context of a rule ('I am castling'), give the outcome ('I killed him'), or provide evaluation ('I did the right thing').

According to Kenny, as we just noted, action sentences have the form 'Jones brought it about that p.' The sentence that replaces 'p' is to be in the present tense, and it describes the result that the agent has wrought: it is a sentence 'newly true of the patient'.[4] Thus, 'The doctor removed the patient's appendix' must be rendered, 'The doctor brought it about that the patient has no appendix.' By insisting that the sentence that replaces 'p' describe a terminal *state* rather than an *event*, it may be thought that Kenny can avoid the criticism made above that the problem of the logical form of action sentences turns up within the sentence that replaces 'p': we may allow that 'The patient has no appendix' presents no relevant problem. The difficulty is that neither will the analysis stand in its present form. The doctor may bring it about that the patient has no

---

[4] Kenny, op. cit., 181.

appendix by turning the patient over to another doctor who performs the operation; or by running the patient down with his Lincoln Continental. In neither case would we say the doctor removed the patient's appendix. Closer approximations to a correct analysis might be, 'The doctor brought it about that the doctor has removed the patient's appendix' or perhaps, 'The doctor brought it about that the patient has had his appendix removed by the doctor.' One may still have a few doubts, I think, as to whether these sentences have the same truth conditions as 'The doctor removed the patient's appendix.' But in any case it is plain that in these versions, the problem of the logical form of action sentences does turn up in the sentences that replace '*p*': 'The patient has had his appendix removed by the doctor' or 'The doctor has removed the patient's appendix' are surely no *easier* to analyse than 'The doctor removed the patient's appendix.' By the same token, 'Cass walked to the store' can't be given as 'Cass brought it about that Cass is at the store', since this drops the idea of walking. Nor is it clear that 'Cass brought it about that Cass is at the store and is there through having walked' will serve; but in any case, the contained sentence is again worse than what we started with.

It is not easy to decide what to do with 'Smith coughed.' Should we say 'Smith brought it about that Smith is in a state of just having coughed'? At best this would be correct only if Smith coughed on purpose.

The difficulty in Kenny's proposal that we have been discussing may perhaps be put this way: he wants to represent every (completed) action in terms only of the agent, the notion of bringing it about that a state of affairs obtains, and the state of affairs brought about by the agent. But many action sentences yield no description of the state of affairs brought about by the action except that it *is* the state of affairs brought about by that action. A natural move, then, is to allow that the sentence that replaces '*p*' in '*x* brings it about that *p*' may (or perhaps must) describe an event.

If I am not mistaken, Chisholm has suggested an analysis that at least permits the sentence that replaces '*p*' to describe (as we are allowing ourselves to say) an event.[5] His favoured locution is '*x* makes *p* happen', though he uses such variants as '*x* brings it about that *p*' or '*x* makes it true that *p*'. Chisholm speaks of the entities to

---

[5] Roderick Chisholm, 'The Descriptive Element in the Concept of Action'. Also see Chisholm, 'The Ethics of Requirement'.

which the expressions that replace '*p*' refer as 'states of affairs', and explicitly adds that states of affairs may be changes or events (as well as 'unchanges'). An example Chisholm provides is this: if a man raises his arm, then we may say he makes it happen that his arm goes up. I do not know whether Chisholm would propose 'Jones made it happen that Jones's arm went up' as an analysis of 'Jones raised his arm', but I think the proposal would be wrong because although the second of these sentences does perhaps entail the first, the first does not entail the second. The point is even clearer if we take as our example 'Jones made it happen that Jones batted an eyelash' (or some trivial variant), and this cannot be called progress in uncovering the logical form of 'Jones batted an eyelash.'

There is something else that may puzzle us about Chisholm's analysis of action sentences, and it is independent of the question what sentence we substitute for '*p*'. Whatever we put for '*p*', we are to interpret it as describing some event. It is natural to say, I think, that *whole* sentences of the form '*x* makes it happen that *p*' also describe events. Should we say that these events are the *same* event, or that they are different? If they are the same event, as many people would claim (perhaps including Chisholm), then no matter what we put for '*p*', we cannot have solved the *general* problem of the logical form of sentences about actions until we have dealt with the sentences that can replace '*p*'. If they are different events, we must ask how the element of agency has been introduced into the larger sentence though it is lacking in the sentence for which '*p*' stands; for each has the agent as its subject. The answer Chisholm gives, I think, is that the special notion of making it happen that he has in mind is intentional, and thus to be distinguished from simply causing something to happen. Suppose we want to say that Alice broke the mirror without implying that she did it intentionally. Then Chisholm's special idiom is not called for; but we could say, 'Alice caused it to happen that the mirror broke.' Suppose we now want to add that she did it intentionally. Then the Chisholm-sentence would be: 'Alice made it happen that Alice caused it to happen that the mirror broke.' And now we want to know, what is the event that the whole sentence reports, and that the contained sentence does not? It is, apparently, just what used to be called an act of the will. I will not dredge up the standard objections to the view that acts of the will are special events distinct from, say, our bodily movements, and perhaps the causes of them. But even if Chisholm is willing to accept such a view, the problem of the logical form of the sentences that can replace '*p*' remains, and these describe the things people do as we describe them when we do not impute intention.

A somewhat different view has been developed with care and precision by von Wright.[6] In effect, von Wright puts action sentences into the following form: '*x* brings it about that a state where *p* changes into a state where *q*'. Thus the important relevant difference between von Wright's analysis and the ones we have been considering is the more complex structure of the description of the change or event the agent brings about: where Kenny and Chisholm were content to describe the result of the change, von Wright includes also a description of the initial state.

Von Wright is interested in exploring the logic of change and action and not, at least primarily, in giving the logical form of our common sentences about acts or events. For the purposes of his study, it may be very fruitful to think of events as ordered pairs of states. But I think it is also fairly obvious that this does not give us a standard way of translating or representing the form of most sentences about acts and events. If I walk from San Francisco to Pittsburgh, for example, my initial state is that I am in San Francisco and my terminal state is that I am in Pittsburgh; but the same is more pleasantly true if I fly. Of course, we may describe the terminal state as my having walked to Pittsburgh from San Francisco, but then we no longer need the separate statement of the initial state. Indeed, viewed as an analysis of ordinary sentences about actions, von Wright's proposal seems subject to all the difficulties I have already outlined plus the extra one that most action sentences do not yield a non-trivial description of the initial state (try 'He circled the field', 'He recited the *Odyssey*', 'He flirted with Olga').

In two matters, however, it seems to me von Wright suggests important and valuable changes in the pattern of analysis we have been considering, or at least in our interpretation of it. First, he says that an action is not an event, but rather the bringing about of an event. I do not think this can be correct. If I fall down, this is an event whether I do it intentionally or not. If you thought my falling was an accident and later discovered I did it on purpose, you would not be tempted to withdraw your claim that you had witnessed an event. I take von Wright's refusal to call an action an event to reflect

[6] Georg Henrik von Wright, *Norm and Action*.

the embarrassment we found to follow if we say that an act is an event, taking agency to be introduced by a phrase like 'brings it about that'. The solution lies, however, not in distinguishing acts from events, but in finding a different logical form for action sentences. The second important idea von Wright introduces comes in the context of his distinction between *generic* and *individual* propositions about events.[7] The distinction, as von Wright makes it, is not quite clear, for he says both: that an individual proposition differs from a generic one in having a uniquely determined truth value, while a generic proposition has a truth value only when coupled with an occasion; and that, that Brutus killed Caesar is an individual proposition while that Brutus kissed Caesar is a generic proposition, because 'a person can be kissed by another on more than one occasion'. In fact the proposition that Brutus kissed Caesar seems to have a uniquely determined truth value in the same sense that the proposition that Brutus killed Caesar does. But it is, I believe, a very important observation that 'Brutus kissed Caesar' does not, by virtue of its meaning alone, describe a single act.

It is easy to see that the proposals we have been considering concerning the logical form of action sentences do not yield solutions to the problems with which we began. I have already pointed out that Kenny's problem, that verbs of action apparently have 'variable polyadicity', arises within the sentences that can replace '*p*' in such formulas as '*x* brought it about that *p*'. An analogous remark goes for von Wright's more elaborate formula. The other main problem may be put as that of assigning a logical form to action sentences that will justify claims that two sentences describe 'the same action'. Our study of some of the ways in which we excuse, or attempt to excuse, acts shows that we want to make inferences such as this: I flew my spaceship to the Morning Star, the Morning Star is identical with the Evening Star; so, I flew my spaceship to the Evening Star. (My leader told me not to go the Evening Star; I headed for the Morning Star not knowing.) But suppose we translate the action sentences along the lines suggested by Kenny or Chisholm or von Wright. Then we have something like, 'I brought it about that my spaceship is on the Morning Star.' How can we infer, given the well-known identity, 'I brought it about that my spaceship is on the Evening Star'? We know that if we replace 'the Morning

[7] Op. cit., 23.

Star' by 'the Evening Star' in, 'My spaceship is on the Morning Star'
the truth-value will not be disturbed; and so if the occurrence of this
sentence in, 'I brought it about that my spaceship is on the Morning
Star' is truth-functional, the inference is justified. But of course the
occurrence can't be truth-functional: otherwise, from the fact that I
brought about one actual state of affairs it would follow that I
brought about every actual state of affairs. It is no good saying that
after the words 'bring it about that' sentences describe something
between truth-values and propositions, say states of affairs. Such a
claim must be backed by a semantic theory telling us how each
sentence determines the state of affairs it does; otherwise the claim
is empty.

Israel Scheffler has put forward an analysis of sentences about
choice that can be applied without serious modification to sentences
about intentional acts.[8] Scheffler makes no suggestion concerning
action sentences that do not impute intention, and so has no solution to the chief problems I am discussing. Nevertheless, his analysis
has a feature I should like to mention. Scheffler would have us
render, 'Jones intentionally buttered the toast' as, 'Jones made-true
a that Jones-buttered-the-toast inscription.' This cannot, for
reasons I have urged in detail elsewhere,[9] be considered a finally
satisfying form for such sentences because it contains the logically
unstructured predicate 'is a that Jones-buttered-the-toast inscription', and there are an infinite number of such semantical primitives
in the language. But in one respect, I believe Scheffler's analysis is
clearly superior to the others, for it implies that introducing the
element of intentionality does not call for a reduction in the content
of the sentence that expresses *what* was done intentionally. This
brings out a fact otherwise suppressed, that, to use our example,
'Jones' turns up twice, once inside and once outside the scope of the
intensional operator. I shall return to this point.

A discussion of the logical form of action sentences in ordinary
language is to be found in the justly famed Chapter VII of Reichenbach's *Elements off Symbolic Logic*.[10] According to Reichenbach's
doctrine, we may transform a sentence like

(4) Amundsen flew to the North pole

[8] Israel Scheffler, *The Anatomy of Inquiry*, 104–5.
[9] Donald Davidson, 'Theories of Meaning and Learnable Languages', 390–1.
[10] Hans Reichenbach, *Elements of Symbolic Logic*, sect. 48.

into:

(5) $(\exists x)$ (x consists in the fact that Amundsen flew to the North Pole).

The expression 'is an event that consists in the fact that' is to be viewed as an operator which, when prefixed to a sentence, forms a predicate of events. Reichenbach does not think of (5) as showing or revealing the logical form of (4), for he thinks (4) is unproblematic. Rather he says (5) is logically equivalent to (4). (5) has its counterpart in a more ordinary idiom:

(6) A flight by Amundsen to the North Pole took place.

Thus Reichenbach seems to hold that we have two ways of expressing the same idea, (4) and (6); they have quite different logical forms, but they are logically equivalent; one speaks literally of events while the other does not. I believe this view spoils much of the merit in Reichenbach's proposal, and that we must abandon the idea that (4) has an unproblematic logical form distinct from that of (5) or (6). Following Reichenbach's formula for putting any action sentence into the form of (5) we translate

(7) Amunsden flew to the North Pole in May 1926

into:

(8) $(\exists x)$ (x consists in the fact that Amundsen flew to the North Pole in May 1926).

The fact that (8) entails (5) is no more obvious than that (7) entails (4); what was obscure remains obscure. The correct way to render (7) is:

(9) $(\exists x)$ (x consists in the fact that Amundsen flew to the North Pole and x took place in May 1926).

But (9) does not bear the simple relation to the standard way of interpreting (7) that (8) does. We do not know of any logical operation on (7) as it would usually be formalised (with a three-place predicate) that would make it logically equivalent to (9). This is why I suggest that we treat (9) alone as giving the logical form of (7). If we follow this strategy, Kenny's problem of the 'variable polyadicity' of action verbs is on the way to solution; there is, of course, no variable polyadicity. The problem is solved in the natural way, by introducing events as entities about which an indefinite number of things can be said.

Reichenbach's proposal has another attractive feature: it eliminates a peculiar confusion that seemed to attach to the idea that sentences like (7) 'describe an event'. The difficulty was that one wavered between thinking of the sentence as describing or referring to that one flight Amundsen made in May 1926, or as describing a kind of event, or perhaps as describing (potentially?) several. As von Wright pointed out, any number of events might be described by a sentence like 'Brutus kissed Caesar.' This fog is dispelled in a way I find entirely persuasive by Reichenbach's proposal that ordinary action sentences have, in effect, an existential quantifier binding the action-variable. When we were tempted into thinking a sentence like (7) describes a single event we were misled: it does not describe any event at all. But if (7) is true, then there is an event that makes it true. (This unrecognized element of generality in action sentences is, I think, of the utmost importance in understanding the relation between actions and desires.)

There are two objections to Reichenbach's analysis of action sentences. The first may not be fatal. It is that as matters stand the analysis may be applied to any sentence whatsoever, whether it deals with actions, events, or anything else. Even '2 + 3 = 5' becomes '$(\exists x)$ (x consists in the fact that 2 + 3 = 5)'. Why not say '2 + 3 = 5' does not show its true colours until put through the machine? For that matter, are we finished when we get to the first step? Shouldn't we go on to '$(\exists y)$ (y consists in the fact that $(\exists x)$ (x consists in the fact that 2 + 3 = 5)'? And so on. It isn't clear on what principle the decision to apply the analysis is based.

The second objection is worse. We have:

(10) $(\exists x)$ (x consists in the fact that I flew my spaceship to the Morning Star)

and

(11) the Morning Star = the Evening Star

and we want to make the inference to

(12) $(\exists x)$ (x consists in the fact that I flew my spaceship to the Evening Star).

The likely principle to justify the inference would be:

(13) (x) (x consists in the fact that S $\leftrightarrow$ x consists in the fact that S')

where 'S'' is obtained from 'S' by substituting, in one or more places, a co-referring singular term. It is plausible to add that (13) holds if 'S' and 'S'' are logically equivalent. But (13) and the last assumption lead to trouble. For observing that 'S' is logically equivalent to '$\hat{y}$ $(y = y$ & $S) = \hat{y}$ $(y = y)$' we get

(14) (x) (x consists in the fact that $S \leftrightarrow x$ consists in the fact that $(\hat{y}(y = y$ & S$) = \hat{y}(y = y)))$.

Now suppose 'R' is any sentence materially equivalent to 'S': then '$\hat{y}(y = y$ & S$)$' and '$\hat{y}(y = y$ & R$)$' will refer to the same thing. Substituting in (14) we obtain

(15) (x) (x consists in the fact that $S \leftrightarrow x$ consists in the fact that $(\hat{y}(y = y$ & R$) = \hat{y}(y = y))$),

which leads to

(16) (x) (x consists in the fact that $S \leftrightarrow x$ consists in the fact that R)

when we observe the logical equivalence of 'R' and '$\hat{y}(y = y$ & $R) = \hat{y}(y = y)$'. (16) may be interpreted as saying (considering that the sole assumption is that 'R' and 'S' are materially equivalent) that all events that occur (= all events) are identical. This demonstrates, I think, that Reichenbach's analysis is radically defective.

Now I would like to put forward an analysis of action sentences that seems to me to combine most of the merits of the alternatives already discussed, and to avoid the difficulties. The basic idea is that verbs of action—verbs that say 'what someone did'—should be construed as containing a place, for singular terms or variables, that they do not appear to. For example, we would normally suppose that 'Shem kicked Shaun' consisted in two names and a two-place predicate. I suggest, though, that we think of 'kicked' as a *three*-place predicate, and that the sentence to be given in this form:

(17) $(\exists x)$ (Kicked(Shem, Shaun, x)).

If we try for an English sentence that directly reflects this form, we run into difficulties. 'There is an event x such that x is a kicking of Shaun by Shem' is about the best I can do, but we must remember 'a kicking' is not a singular term. Given this English reading, my proposal may sound very like Reichenbach's; but of course it has quite different logical properties. The *sentence* 'Shem kicked Shaun' nowhere appears inside my analytic sentence, and this makes it differ from all the theories we have considered.

The principles that license the Morning Star-Evening Star inference now make no trouble: they are the usual principles of extensionality. As a result, nothing now stands in the way of giving a standard theory of meaning for action sentences, in the form of a Tarski-type truth definition; nothing stands in the way, that is, of giving a coherent and constructive account of how the meanings (truth conditions) of these sentences depend upon their structure. To see how one of the troublesome inferences now goes through, consider (10) rewritten as

(18) $(\exists x)\,(\mathrm{Flew}(\mathrm{I}, \text{my spaceship}, x)\,\&\,\mathrm{To}(\text{the Morning Star}, x))$.

which, along with (11), entails

(19) $(\exists x)\,(\mathrm{Flew}(\mathrm{I}, \text{my spaceship}, x)\,\&\,\mathrm{To}(\text{the Evening Star}, x))$.

It is not necessary, in representing this argument, to separate off the To-relation; instead we could have taken, 'Flew' as a four-place predicate. But that would have obscured *another* inference, namely that from (19) to

(20) $(\exists x)\,(\mathrm{Flew}(\mathrm{I}, \text{my spaceship}, x))$.

In general, we conceal logical structure when we treat prepositions as integral parts of verbs; it is a merit of the present proposal that it suggests a way of treating prepositions as contributing structure. Not only is it good to have the inference from (19) to (20); it is also good to be able to keep track of the common element in 'fly to' and 'fly away from' and this of course we cannot do if we treat these as unstructured predicates.

The problem that threatened in Reichenbach's analysis, that there seemed no clear principle on which to refrain from applying the analysis to every sentence, has a natural solution if my suggestion is accepted. Part of what we must learn when we learn the meaning of any predicate is how many places it has, and what sorts of entities the variables that hold these places range over. Some predicates have an event-place, some do not.

In general, what kinds of predicates do have event-places? Without pursuing this question very far, I think it is evident that if action predicates do, many predicates that have little relation to action do. Indeed, the problems we have been mainly concerned with are not at all unique to talk of actions: they are common to talk of events of any kind. An action of flying to the Morning Star is identical with an action of flying to the Evening Star; but equally, an eclipse of the Morning Star is an eclipse of the Evening Star. Our ordinary talk of events, of causes and effects, requires constant use of the idea of different descriptions of the same event. When it is pointed out that striking the match was not sufficient to light it, what is not sufficient is not the event, but the description of it—it was a *dry* match, and so on.[11] And of course Kenny's problem of 'variable polyadicity', though he takes it to be a mark of verbs of action, is common to all verbs that describe events.

It may now appear that the apparent success of the analysis proposed here is due to the fact that it has simply omitted what is peculiar to action sentences as contrasted with other sentences about events. But I do not think so. The concept of agency contains two elements, and when we separate them clearly, I think we shall see that the present analysis has not left anything out. The first of these two elements we try, rather feebly, to elicit by saying that the agent acts, or does something, instead of being acted upon, or having something happen to him. Or we say that the agent is active rather than passive; and perhaps try to make use of the moods of the verb as a grammatical clue. And we may try to depend upon some fixed phrase like 'brings it about that' or 'makes it the case that'. But only a little thought will make it clear that there is no satisfactory grammatical test for verbs where we want to say there is agency. Perhaps it is a *necessary* condition of attributing agency that one argument-place in the verb is filled with a reference to the agent as a person; it will not do to refer to his body, or his members, or to anyone else. But beyond that it is hard to go. I sleep, I snore, I push buttons, I recite verses, I catch cold. Also others are insulted by me, struck by me, admired by me, and so on. No grammatical test I know of, in terms of the things we may be said to do, of active or passive mood, or of any other sort, will separate out the cases here where we want to speak of agency. Perhaps it is true that 'brings it about that' guarantees agency; but as we have seen, many sentences that do attribute agency cannot be cast in this grammatical form.

[11] See Essay 7 for more on this topic.

I believe the correct thing to say about *this* element in the concept of agency is that it is simply introduced by certain verbs and not by others; when we understand the verb we recognize whether or not it includes the idea of an agent. Thus, 'I fought' and 'I insulted him' do impute agency to the person referred to by the first singular term, 'I caught cold' and, 'I had my thirteenth birthday' do not. In these cases, we do seem to have the following test: we impute agency only where it makes sense to ask whether the agent acted intentionally. But there are other cases, or so it seems to me, where we impute agency only when the answer to the question whether the agent acted intentionally is 'yes'. If a man falls down by accident or because a truck knocks him down, we do not impute agency; but we do if he fell down on purpose.[12]

This introduces the second element in the concept of agency, for we surely impute agency when we say or imply that the act is intentional. Instead of speaking of two elements in the concept of agency, perhaps it would be better to say there are two ways we can imply that a person acted as an agent: we may use a verb that implies it directly, or we may use a verb that is non-committal, and add that the act was intentional. But when we take the second course, it is important not to think of the intentionality as adding an extra doing of the agent; we must not make the expression that introduces intention a verb of action. In particular, we cannot use 'intentionally brings it about that' as the expression that introduces intention, for 'brings it about that' is in itself a verb of action, and imputes agency, but it is neutral with respect to the question whether the action was intentional as described.

This leaves the question what logical form the expression that introduces intention should have. It is obvious, I hope, that the adverbial form must be in some way deceptive; intentional actions are not a class of actions, or, to put the point a little differently, doing something intentionally is not a manner of doing it. To say someone did something intentionally is to describe the action in a way that bears a special relation to the beliefs and attitudes of the agent; and perhaps further to describe the action as having been caused by those beliefs and attitudes.[13] But of course to describe the action of the agent as having been caused in a certain way does not mean that the agent is described as performing any further action.

[12] See Essay 3.　　　[13] See Essay 1.

From a logical point of view, there are thus these important conditions governing the expression that introduces intention: it must not be interpreted as a verb of action, it must be intensional, and the intention must be tied to a person. I propose then that we use some form of words like 'It was intentional of x that p' where 'x' names the agent, and 'p' is a sentence that says the agent did something. It is useful, perhaps necessary, that the agent be named twice when we try to make logical form explicit. It is useful, because it reminds us that to describe an action as intentional is to describe the action in the light of certain attitudes and beliefs of a particular person; it may be necessary in order to illuminate what goes on in those cases in which the agent makes a mistake about who he is. It was intentional of Oedipus, and hence of the slayer of Laius, that Oedipus sought the slayer of Laius, but it was not intentional of Oedipus (the slayer of Laius) that the slayer of Laius sought the slayer of Laius.

CRITICISM, COMMENT, AND DEFENCE

Essay 6 brought in its wake a number of comments and criticisms from other philosophers, and in a few cases I responded. In this appendix to Essay 6 I bring together some of my replies, for although they repeat much that can be found elsewhere in this volume, they often put a point in a new way or modify an old one. I have done some editing to make these replies intelligible without the comments to which they were replies, but of course some readers may want to look up the original work of the critic or commentator.

This Essay was first read at a three-day conference on *The Logic of Decision and Action* held at the University of Pittsburgh in March 1966; the proceedings were published the next year under the editorship of Nicholas Rescher. At the conference, E. J. Lemmon, H.-N. Castañeda, and R. M. Chisholm commented on my paper, and I replied. It is my replies (as rewritten for publication) that appear here (somewhat further edited).

In November of 1966 The University of Western Ontario held a colloquium on *Fact and Existence* at which I replied to a paper 'On Events and Event-Descriptions' by R. M. Martin. Both his paper and my reply were published by Blackwells in 1969 under the editorship of Joseph Margolis. Martin had not seen my Essay 6 when he wrote his paper, and in fairness to him it should be noted that his views on the semantics of sentences about events have been modified subsequently. I reprint my reply to him for the light it throws on my views, not on his.

Finally, the journal *Inquiry* devoted its Summer, 1970, issue to the subject of action, and it contained two criticisms of my work. One was by

Carl G. Hedman, 'On the Individuation of Actions', the other was by James Cargile, 'Davidson's Notion of Logical Form'. My replies were printed under the title 'Action and Reaction', and are reprinted here.

A. *Reply to Lemmon on Tenses*. My goal was to get clear about the logical form of action sentences. By action sentences I mean sentences *in English* about actions. At the level of abstraction on which the discussion moved, little was said that would not apply to sentences about actions in many other languages if it applied to sentences in English. The ideal implicit in the paper is a theory that spells out every element of logical form in every English sentence about actions. I dream of a theory that makes the transition from the ordinary idiom to canonical notation purely mechanical, and a canonical notation rich enough to capture, in its dull and explicit way, every difference and connection legitimately considered the business of a theory of meaning. The point of canonical notation so conceived is not to improve on something left vague and defective in natural language, but to help elicit in a perspicuous and general form the understanding of logical grammar we all have that constitutes (part of) our grasp of our native tongue.

In exploring the logical form of sentences about actions and events, I concentrated on certain features of such sentences and neglected others. One feature I totally neglected was that of tense; Lemmon is absolutely right in pointing out that some of the inferences I consider valid depend (in a standard way we have become hardened to) on fudging with respect to time. The necessity for fudging shows that we have failed to bring out a feature of logical form.

I accept the implication that my own account was incomplete through neglect of the element of tense, and I welcome Lemmon's attempt to remedy the situation. I am very much in sympathy with the methods he apparently thinks appropriate. Logicians have almost always assumed that the demonstrative element in natural languages necessarily resists serious semantic treatment, and they have accordingly tried to show how to replace tensed expressions with others containing no demonstrative feature. What recommends this strategy to logicians (the elimination of sentences with variable truth-values) also serves to show that it is not a strategy for analysing the sentences of English. Lemmon makes no attempt to eliminate the demonstrative element from his canonical notation (substituting 'before now' for the past tense is a way of *articulating* the relation between the different tenses of the same verb, not of eliminating the demonstrative element). At the same time, he obviously has in mind that the structure he introduces must lend itself to formal semantic treatment. It is simply a mistake, Lemmon correctly assumes, to think that sentences with a demonstrative element resist the application of systematic semantic analysis.

B. *Reply to Lemmon on Identity Conditions for Events.* If we are going to quantify over events and interpret singular terms as referring to events, we need to say something about the conditions under which expressions of the form '$a = b$' are true where 'a' and 'b' refer, or purport to refer, to events. This is a difficult and complex subject, and I do not propose to do more here than comment briefly on some of Lemmon's remarks. But I think he is right to raise the issue; before we decide that our general approach to the analysis of event sentences is correct, there must be much more discussion of the criteria for individuating and identifying events.

   Lemmon is surely right that a necessary condition for the identity of events is that they take place over exactly the same period of time. He suggests, very tentatively, that if we add that the events 'take the same place', then we have necessary and sufficient conditions for identity. I am not at all certain this suggestion is wrong, but before we accept it we shall need to remove two doubts. The first centres on the question whether we have adequate criteria for the location of an event. As Lemmon realizes, his principle that if F (a,z) then a is a participant in z, cannot be true for every F (take 'F' as 'took place a thousand miles south of' and 'a' as 'New York'; we would not, presumably, say New York participated in every event that took place a thousand miles south of New York). And how do we deal with examples like this: if a man's arm goes up, the event takes place in the space-time zone occupied by the arm; but if a man raises his arm, doesn't the event fill the zone occupied by the whole man? Yet the events may be identical. If a man drives his car into his garage, how much of the garage does the event occupy? All of it, or only the zone occupied by the car? Finally, if events are to have a location in an interesting sense, we need to see what is wrong with the following argument: if an event is a change in a certain object, then the event occupies at least the zone occupied by the object during the time the event takes place. But if one object is part of another, a change in the first is a change in the second. Since an object is part of the universe, it follows that every event that is a change in an object takes place everywhere (throughout the universe). This argument is, I believe, faulty, but it must be shown to be so before we can talk intelligibly of the location of events.

The second doubt we must remove if we are to identify events with space-time zones is that there may be two different events in the same zone. Suppose that during exactly the same time interval Jones catches cold, swims the Hellespont, and counts his blessings. Are these all the same event? I suspect there may be a good argument to show they are; but until one is produced, we must suspend judgement on Lemmon's interesting proposal.[14]

C. *Reply to Castañeda on Agent and Patient.* Castañeda very usefully summarizes the main points in my paper, and raises some questions about the principles that are implicit in my examples. My lack of explicitness has perhaps misled him in one respect. It is not part of my programme to make all entailments matters of logical form. '$x > y$' entails '$y < x$', but not as a matter of form. 'x is a grandfather' entails 'x is a father', but not as a matter of form. And I think there are cases where, to use Castañeda's words, 'a larger polyadic action statement entails a shorter one which is a part of it' and yet this is not a matter of logical form. An example, perhaps, is this: 'I flew my spaceship' may entail, 'I flew', but if it does, it is not, I think, because of the logical form of the sentences. My reason for saying this is that I find no reason to believe the logical form of 'I flew my spaceship' differs from that of 'I sank the *Bismarck*', which does not entail 'I sank' though it does happen to entail 'The *Bismarck* sank'. A comparison of these examples ought to go a long way to persuade us that simple sentences containing transitive verbs do not, as a matter of logical form, entail sentences with intransitive verbs. Putting sentences in the passive will not radically change things. If I sank the *Bismarck*, the *Bismarck* was sunk and the *Bismarck* sank. But 'The *Bismarck* was sunk' and 'The *Bismarck* sank' are not equivalent, for the second does not entail the first. Thus even if we were to accept Castañeda's view that 'The *Bismarck* was sunk' has a logically intransitive verb, the passivity of the subject remains a feature of this verb distinguishing it from the verb

[14] For more on the individuation of events, see Essay 8.

of 'The *Bismarck* sank'. Thus there is no obvious economy in Castañeda's idea of indicating the distinction between agent and patient by position in verbs of action. There would be real merit, however, in keeping track of the relation between 'The *Bismarck* was sunk' and 'The *Bismarck* sank', which is that the first entails the second; but Castañeda's notation does not help with this.

Castañeda would have us put 'The King insulted the Queen' in this form:

$$(\exists x) \, (\text{Insulted (the King, } x) \ \& \ \text{Insulted } (x, \text{ the Queen}))$$

What is this relation, that relates a person and an event or, *in the same way*, an event and a person? What logical feature is preserved by this form that is not as well preserved, and less misleadingly, by

$$(\exists x) \, (\text{Insulted (the King, } x) \ \& \ \text{Was insulted (the Queen, } x))$$

(i.e., 'There was an event that was an insulting by the King and of the Queen')? But I remain unconvinced of the advantages in splitting transitive verbs up in this way. The gain is the entailment of 'My spaceship was flown' by 'I flew my spaceship'; the loss becomes apparent when we realize that 'My spaceship was flown' has been interpreted so as not to entail 'Someone flew my spaceship'.[15]

D. *Reply to Castañeda on Prepositions.* My proposal to treat certain prepositions as verbs does seem odd, and perhaps it will turn out to be radically mistaken. But I am not quite convinced of this by what Castañeda says. My analysis of 'I flew my spaceship to the Morning Star' does entail '$(\exists x) \, (\text{To (the Morning Star, } x))$', and Castañeda turns this into words as 'There was a to the Morning Star'. But I think we can do better: 'There was an event involving motion toward the Morning Star' or 'There was an event characterized by being to (toward) the Morning Star'. Castañeda himself proposes 'flying-to', which shows he understands the *sort* of verb I have in mind. But of course I don't like 'flying-to' as an unstructured predicate, since this breaks the connection with 'walking-to' and its kin. Castañeda complains, of my use of plain 'to', that there are many different senses of 'to', depending on the verb it is coupled with. Let us suppose we understand this difficulty, with its heavy

[15] On the general point raised by Castañeda, whether transitive verbs entail their intransitive counterparts as a matter of logical form, and (a related matter) whether passive transformation is a matter of logical form, I would now side with Castañeda.

dependence on the concept of sameness of relation. I shall meet Castañeda half-way by introducing a special form of 'to' which means, 'motion-toward-and-terminating-at'; this is more general than his 'flying-to' and less general than my former, plain, 'to'. And I assume that if Castañeda understands '($\exists x$) (flying-to (the Morning Star, x))' he will understand '($\exists x$) (Motion-towards-and-terminating-at (the Morning Star, x))', for this verb differs from his merely in degree of generality.

E. *Reply to Castañeda on Intention.* First Castañeda makes the point, also made by Lemmon, that I would have done well to make basic a notion of intention that does not imply that what is intended is done. I think they are right in this.

Castañeda then goes on to claim that my analysis of 'Oedipus intentionally sought the slayer of Laius' as 'It was intentional of Oedipus that Oedipus sought the slayer of Laius' is faulty because the first sentence might be true and the second false if Oedipus failed to know that he was Oedipus. Castañeda suggests that to correct the analysis, we should put 'he (himself)' for the second occurrence of 'Oedipus'. In my opinion, Castañeda is right both in his criticism and in his correction. There is, as he maintains, an irreducibly demonstrative element in the full analysis of sentences about intentions, and my proposal concealed it.

Perhaps I should remark here that I do not think it *solves* the problem of the analysis of sentences about intention to put them in the form of 'It was intentional of x that p'; such sentences are notoriously hard to bring under a semantical theory. I view putting such sentences in this form as a first step; the problem then looks, even with Castañeda's revision, much like the problem of analysing sentences about other propositional attitudes.

F. *Reply to Chisholm on Making Happen.* I am happy to have Chisholm's careful comments on the section of my paper that deals with his views; he has made me realize that I had not appreciated the subtlety of his analysis. It is not clear to me now whether, on the issues discussed in my paper, there is any disagreement between us. Let me formulate the questions that remain in my mind as I now understand them.

I assume that since he has not attempted an analysis of event sentences generally, and the '*p*' in, 'He made it happen that *p*' refers to an event, Chisholm does not dispute my claim that he has not solved the main problems with which I deal in my paper. The question is rather whether there are any special problems in his analysis of action and agency. The first difficulty I raised for Chisholm was whether he could produce, in a reasonably mechanical way, for every sentence of the form 'He raised his arm' or 'Alice broke the mirror', another sentence of the form 'He made it happen that *p*' or 'Alice made it happen that *p*' where '*p*' does not have the agent as subject. Chisholm shows, I think, that there is a chance he can handle 'He raised his arm' and 'Alice broke the mirror' except, perhaps, in the case where intention is not involved at all, and this is not under discussion. The cases I would now worry about are rather 'He walked to the corner', 'He carved the roast', 'He fell down', or 'The doctor removed the patient's appendix'. In each of these examples I find I am puzzled as to what the agent makes happen. My problem isn't that I can't imagine that there is some bodily movement that the agent might be said to make happen, but that I see no way *automatically* to produce the right description from the original sentence. No doubt each time a man walks to the corner there is some way he makes his body move; but of course it does not follow that there is some one way he makes his body move every time he walks to the corner.

The second difficulty I raised for Chisholm concerned the question whether his analysis committed him to 'acts of the will', perhaps contrary to his own intentions. It is clear that Chisholm does not *want* to be committed to acts of the will, and that his analysis does not *say* that there are acts of the will but I believe the question can still be raised. It can be raised by asking whether the event said to occur in 'Jones made it happen that his arm went up' is the same event or a different one from the event said to occur in 'Jones's arm went up'. It seems to me Chisholm can avoid acts of the will only by saying the events are the same. He is free to say this, of course, and then the only objection is terminological. And 'Jones's arm went up' would then be, when it was something Jones made happen, a description of an action.

At the end of his reply, Chisholm conjectures that I may not agree with him that agents may be causes. Actually I see no objection to saying that agents are causes, but I think we understand this only when we can reduce it to the case of an event being a cause; and here

I do disagree with Chisholm.[16] He asks how we are to render 'He made it happen that *p*' in terms merely of relations among events. If the problem is that of giving the logical form of action sentences, then I have made a suggestion in the present paper. If the problem is to give an *analysis* of the concept of agency using other concepts, then I am not sure it can be done. Why must it be possible?

G. *Reply to Martin*. There is a more or less innocent sense in which we say that a sentence refers to, describes, or is about, some entity when the sentence contains a singular term that refers to that entity. Speaking in this vein, we declare that, 'The cat has mange' refers to the cat, 'Caesar's death was brought on by a cold' describes Caesar and his death, and 'Jack fell down and broke his crown' is about Jack and Jack's crown. Observing how the reference of a complex singular term like 'Caesar's death' or 'Jack's crown' depends systematically on the reference of the contained singular term ('Caesar' or 'Jack') it is tempting to go on to ask what a sentence *as a whole* is about (or refers to, or describes), since it embraces singular terms like 'Caesar's death' in much the way 'Caesar's death' embraces 'Caesar'. There is now a danger of ambiguity in the phrases 'what a sentence refers to' or 'what a sentence is about'; let us resolve it by using only 'refers to' for the relation between patent singular terms and what they are about, and only 'corresponds to' for the relation between a sentence and what it is about.

Just as a complex singular term like 'Caesar's death' may fail of reference though contained singular terms do not, so a sentence may not correspond to anything, even though its contained singular terms refer; witness 'Caesar's death was brought on by a cold'. Clearly enough, it is just the true sentences that have a corresponding entity; 'The cat has mange' corresponds to the cat's having of mange, which alone can make it true; because there is no entity that is Caesar's death having been brought on by a cold, 'Caesar's death was brought on by a cold' is not true.[17]

---

[16] See Essay 3.

[17] For simplicity's sake I speak as if truth were a property of sentences: more properly it is a relation between a sentence, a person and a time. (We could equally think of truth as a property of utterances, of tokens, or of speech acts.) I assume here that when truth is attributed to a sentence, or reference to a singular term, the suppressed relativization to a speaker and a time could always be supplied; if so, the ellipsis is harmless.

These gerunds can get to be a bore, and we have a way around them in 'fact that' clauses. The entity to which 'The cat has mange' corresponds is the cat's having of mange; equivalently, it is the fact that the cat has mange. Quite generally we get a singular term for the entity to which a sentence corresponds by prefixing 'the fact that' to the sentence; assuming, of course, there are such entities.

Philosophical interest in facts springs partly from their promise for explaining truth. It's clear that most sentences would not have the truth value they do if the world were not the way it is, but *what* in the world makes a sentence true? Not just the objects to which a sentence refers (in the sense explained above), but rather the doings and havings of relations and properties of those objects; in two words, the facts. It seems that a fact contains, in appropriate array, just the objects any sentence it verifies is about. No wonder we may not be satisfied with the colourless 'corresponds to' for the relation between a true sentence and its fact; there is something, we may feel, to be said for 'is true to', 'is faithful to', or even 'pictures'.

To specify a fact is, then, a way of explaining what makes a sentence true. On the other hand, simply to say that a sentence is true is to say there is some fact or other to which it corresponds. On this account, '*s* is true to (or corresponds to) the facts' means more literally '*s* corresponds to a fact'. Just as we can say there is a fact to which a sentence corresponds when the sentence is true, we can also say there is a true sentence corresponding to a particular fact; this latter comes down to saying of the fact that it is one. English sentences that perhaps express this idea are 'That the cat has mange is a fact' and 'It is a fact that London is in Canada', and even 'London is in Canada, and that's a fact.' It is evident that we must distinguish here between idioms of at least two sorts, those that attribute facthood to an entity (a fact), and those that say of a sentence that it corresponds to a fact (or 'the facts'). Let us use the following sentences as our samples of the two sorts of idiom:

(1) That the cat has mange is a fact.
(2) The sentence, 'The cat has mange' corresponds to a fact.

Professor Martin says his analysis is intended to apply to sentences of the form 'So-and-so is a fact' where I suppose 'so-and-so' is to be replaced, typically, by a that-clause, and he suggests we interpret such sentences as saying of a sentence that it is true (non-analytically—but I shall ignore this twist). Which of the two idioms represented by (1) and (2) is Martin analysing? The sentences Martin says he wants to analyse apparently have the form of (1); his analysis, on the other hand, seems suited to sentences like (2).

Suppose we try the second tack. Then Martin's proposal comes to this: where we appear to say of a sentence that there is a fact to which it corresponds we might as well say simply that the sentence is true. There is nothing in this yet to offend the most devoted friend of facts. Martin has not explained away a singular term that ever purported to refer to a fact; on his analysis, as on the one the friend of facts would give, the only singular term in, 'The sentence "The cat has mange" corresponds to the facts' refers to a sentence. Nor would the friend of facts want to deny the equivalence of '*s* is true' and '*s* corresponds to a fact' when '*s*' is replaced by the name or description of a sentence. The friend of facts would, however, balk at the claim that this shows how, *in general*, to eliminate quantification over facts, or singular terms that refer to them. He would contend that it is only sentence (1) with its apparent singular term 'that the cat has mange' which clearly calls for an ontology of facts. Martin may reply that it is sentence (1) he had his eye on from the start. This reply leaves (2) out in the cold unless, of course, (1) and (2) can be given the same analysis. The partisan of facts will resist this idea, and plausibly, I think, on the ground that (2) is merely an existential generalization of the more interesting:

> (3) The sentence, 'The cat has mange' corresponds to the fact that the cat has mange.

Here Martin's attempt to treat facts as sentences cannot be made to work without reducing (3) to the statement that the sentence, 'The cat has mange' corresponds to itself, and this cannot be right since (3), like (2), is clearly *semantical* in character; it relates a sentence to the world. Martin recognizes the semantic thrust in talk of facts, but does not notice that it cannot be reconciled with his analysis of (1).

Martin's thesis that we do not need an ontology of facts could still be saved by an argument to show that there is at most one fact, for the interest in taking sentences like (3) as containing singular terms referring to facts depends on the assumption that there is an indefinitely large number of different facts to be referred to: if there were only one, we could submerge reference to it into what might as well be considered a one-place predicate.[18] And an argument is handy, thanks to Frege, showing that if sentences refer at all, all true sentences must refer to the same thing.[19]

We may then with easy conscience side with Martin in viewing 'corresponds to a fact', when said of a sentence, as conveying no more than 'is true'. What should we say of the sentences like (1) that appear to attribute facthood to entities? As we have seen, such sentences cannot be analysed as being about sentences. Bearing in mind the unity of fact, we might say (1) affirms The Great Fact, or tells The Truth, by way of one of its infinity of tags, 'The cat has mange.' We could equally well accept the universe with 'That London is in Canada is a fact.' Equivalently, we could have simply said, 'London is in Canada.' So, on my account, 'The sentence "The cat has mange" corresponds to the facts' comes out 'The sentence "The cat has mange" is true', but 'That the cat has mange is a fact' comes out just 'The cat has mange'; not at all the same thing.[20]

It is often assumed or argued (though not by Martin) that events are a species of fact. Austin, for example, says, 'Phenomena, events, situations, states of affairs are commonly supposed to be genuinely-in-the-world. . . . Yet surely of all these we can say that they *are facts*. The collapse of the Germans is an event and is a fact—was an event and was a fact'.[21] Reichenbach even treats the words 'event' and 'fact' as synonyms, or so he says.[22] The pressure to treat events as facts is easy, in a way, to understand: both offer themselves as what sentences—some sentence at least—refer to or are about. Causal laws, we are told, say that every event of a certain sort is followed by an event of another sort. According to Hempel, the sentence, 'The length of copper rod r increased between 9.00 and 9.01 a.m.' describes a particular event.[23] In philosophical discussion of action these days we very often learn such things as that 'Jones raised his arm' and 'Jones signalled' may describe the same action, or that an agent may perform an action intentionally under one description

[18] For a more general treatment of 'ontological reduction' by incorporation of a finite number of singular terms into predicates, see Quine's 'Existence and Quantification' and 'Ontological Reduction and the World of Numbers', 203.

[19] For the argument, see Essay 6. For the argument and discussion, see A. Church, *Introduction to Mathematical Logic*, 24–5.

[20] I think that failure to observe the distinction between these two cases is the cause of some of the endless debate whether attributions of truth are redundant.

[21] J. L. Austin, 'Unfair to Facts', 104.

[22] Hans Reichenbach, *Elements of Symbolic Logic*, 269.

[23] Carl Hempel, *Aspects of Scientific Explanation*, 421.

and not under another. It is obvious that most of the sentences usually said to be about events contain no singular terms that even appear to refer to events, nor are they normally shown to have variables that take events as values when put over into ordinary quantificational notation. The natural conclusion is that sentences as wholes describe or refer to events, just as they were said to correspond as wholes to facts, and this, as we have seen, must be wrong.

Martin does not fall into this common trap, for although he constructs singular terms for events from the material of a sentence, he does not have the sentence itself refer to an event. His procedure is to view an event as an ordered n-tuple made up of the extensions of the $n-1$ singular terms and the $n-1$-place predicate of a true sentence. So, 'Leopold met Stephen on Bloomsday' gives us the singular term, '$\langle M, l, s, b \rangle$' which refers to Leopold's meeting of Stephen on Bloomsday provided Leopold did meet Stephen on Bloomsday. I shall ignore the further step by which Martin eliminates ordered n-tuples in favour of virtual ordered n-tuples; the difficulties about to appear are independent of that idea.[24]

Given the premise that Bloomsday is 16 June 1904, we may infer from, 'Leopold met Stephen on Bloomsday' the sentence, 'Leopold met Stephen on 16 June 1904', and, events being the ordered n-tuples they are, Leopold's meeting of Stephen on Bloomsday is identical with Leopold's meeting of Stephen on 16 June 1904. This is surely as it should be so far; but not, I'm afraid, farther. Not every encounter is a meeting; according to the story, some encounters between Leopold and Stephen are meetings and some are not. But then by Martin's account no meeting is identical with an encounter, though between the same individuals and at the same time. The reason is that if any encounter is not a meeting, $\langle E, l, s, b \rangle$ is not identical with $\langle M, l, s, b \rangle$. Indeed, Leopold's first meeting with Stephen on Bloomsday in Dublin cannot be identical with Leopold's first meeting with Stephen on Bloomsday (since a four-place predicate can't have the same extension as a three-place predicate); nor can a meeting between Stephen and Bloom be identical with a meeting between Bloom and Stephen (since entities will be ordered in a different way). No stabbing can be a killing and

---

[24] Substantially the same analysis of events as Martin's has been given by Jaegwon Kim, 'On the Psycho-Physical Identity Theory'. Kim does not take the extra step from real to virtual n-tuples.

no killing can be a murder, no arm-raising a signalling, and no birthday party a celebration. I protest.

Martin's conditions on identity of events are clearly not necessary, but are they perhaps sufficient? Again I think the answer is no. Martin correctly remarks that on his analysis the expressions that are supposed to refer to events refer to one entity at most; but are these entities the events they should be? Suppose Leopold met Stephen more than once on Bloomsday; what unique meeting does Martin's ordered n-tuple pick out? 'Leopold's meeting with Stephen on Bloomsday', like Martin's '$\langle M, l, s, b \rangle$', is a true singular term. But there is this difference, that the first refers to a meeting if it refers to anything, while the second does not. Being more specific about time will not really mend matters: John's kissing of a girl at precisely noon is not a unique kissing if he kissed two girls simultaneously. Martin's method cannot be systematically applied to form singular terms guaranteed to pick out a particular kissing, marriage, or meeting if anything; but this is easy, with gerund phrases, in English.

Martin's mistake is natural, and it is connected with a basic confusion about the relation between a sentence like 'Leopold met Stephen on Bloomsday' or 'Caesar died' and particular events like Leopold's meeting with Stephen on Bloomsday or Caesar's death. The mistake may be encapsulated in the idea (common to Martin and many others) that 'Leopold met Stephen on Bloomsday' comes to the same as 'Leopold's meeting with Stephen on Bloomsday occurred' or that 'Caesar died' may be rendered 'Caesar's death took place'. 'Caesar's death', like 'Leopold's meeting with Stephen', is a true singular term, and so 'Caesar's death took place' and 'Leopold's meeting with Stephen occurred' are true only if there was just one such meeting or death. But 'Caesar died' is true even if Caesar died a thousand deaths, and Leopold and Stephen may meet as often as they please on Bloomsday without falsifying 'Leopold met Stephen on Bloomsday.'

A sentence such as 'Vesuvius erupted in A.D. 79' no more refers to an individual event than 'There's a fly in here' refers to an individual fly. Of course there may be just one eruption that verifies the first sentence and just one fly that verifies the second; but that is beside the point. The point is that neither sentence can properly be interpreted as referring or describing, or being about, a particular eruption or fly. No singular term for such is in the offing. 'There's a fly in here' is existential and general with respect to flies in here;

'Vesuvius erupted in A.D. 79' is existential and general with respect to eruptions of Vesuvius in A.D. 79—if there are such things as eruptions, of course.

Here I am going along with Ramsey who, in a passage quoted by Martin, wrote, '"That Caesar died" is really an existential proposition, asserting the existence of an event of a certain sort, thus resembling "Italy has a King", which asserts the existence of a man of a certain sort. The event which is of that sort is called the death of Caesar, and should no more be confused with the fact that Caesar died than the King of Italy should be confused with the fact that Italy has a King.'[25] This seems to me nearly exactly right: facts, if such there are, correspond to whole sentences, while events, if such there are, correspond to singular terms like 'Caesar's death', and are quantified over in sentences such as 'Caesar died.'[26]

Martin says he doubts that 'Caesar died' must, or perhaps even can, be construed as asserting the existence of an event of a certain sort. I want to demonstrate briefly first that it can, and then, even more briefly, why I think it must.

It can be done by providing event-verbs with one more place than we generally think necessary, a place for events. I propose that 'died' in 'Caesar died' be taken as a two-place predicate, one place for 'Caesar' and another for a variable ranging over events. The sentence as a whole then becomes '$(\exists x)$ (Died (Caesar, x))', that is, there exists a Caesar-dying event, or there exists an event that is a dying of Caesar. There is no problem in forming a singular term like 'Caesar's death' from these materials: it is '(ιx) (Died (Caesar, x))'. We may then say truly, though this is not equivalent to 'Caesar died', that Caesar died just once: '$(\exists y)$ $(y = (\iota x)$ (Died (Caesar, x)))'; we may even say Caesar died Caesar's death: 'Died (Caesar, (ιx) (Died (Caesar, x)))'.

This gives us some idea what it would be like to treat events seriously as individuals, with variables ranging over them, and with corresponding singular terms. It is clear, I think, that none of the

[25] Ramsey, *Foundations of Mathematics*, 138ff.
[26] Austin blundered when he thought a phrase like 'the collapse of the Germans' could unambiguously refer to a fact and to an event. Zeno Vendler very shrewdly uncovers the error, remarking that 'in as much as the collapse of the Germans is a fact, it can be mentioned or denied, it can be unlikely or probable, it can shock or surprise us; in as much as it is an event, however, and not a fact, it can be observed and followed, it can be sudden, violent, or prolonged, it can occur, begin, last and end.' This is from 'Comments' by Vendler (on a paper by Jerrold Katz).

objections I have considered to Reichenbach's, Kim's, or Martin's analyses apply to the present suggestion. We *could* introduce an ontology of events in this way, but of course the question remains whether there is any good reason to do so. I have already mentioned some of the contexts, in the analysis of action, of explanation, and of causality in which we seem to need to talk of events; still, faced with a basic ontological decision, we might well try to explain the need as merely seeming. There remains however a clear problem that is solved by admitting events, and that has no other solution I know of.

The problem is simple, and ubiquitious. It can be illustrated by pointing out that 'Brutus stabbed Caesar in the back in the Forum with a knife' entails 'Brutus stabbed Caesar in the back in the Forum' and both these entail 'Brutus stabbed Caesar in the back' and all these entail 'Brutus stabbed Caesar'; and yet our common way of symbolizing these sentences reveals no logical connection. It may be thought the needed entailments could be supplied by interpreting 'Brutus stabbed Caesar' as elliptical for 'Brutus stabbed Caesar somewhere (in Caesar) somewhere (in the world) with something', but this is a general solution only if we know some fixed number of places for the predicate 'stabbed' large enough to accommodate all eventualities. It's unlikely we shall succeed, for a phrase like 'by' can introduce an indefinitely large number of modifications, as in 'He hung the picture by putting a nail in the wall, which in turn he did by hitting the nail with a hammer, which in turn he did by . . .'.[27] Intuitively, there is no end to what we can say about the causes and consequences of events; our theory of language has gone badly astray if we must treat each adverbial modification as introducing a new place into a predicate. The problem, you can easily persuade yourself, is not peculiar to verbs of action.

My proposed analysis of sentences with event-verbs solves this difficulty, for once we have events to talk about, we can say as much or as little as we please about them. Thus the troublesome sentence becomes (not in symbols, and not quite in English): 'There exists an event that is a stabbing of Caesar by Brutus event, it is an into the back of Caesar event, it took place in the Forum, and Brutus did it with a knife.' The wanted entailments now go through as a matter of form.

Before we enthusiastically embrace an ontology of events we will

---

[27] I am indebted to Daniel Bennett for the example.

want to think long and hard about the critieria for individuating them. I am myself inclined to think we can do as well for events generally as we can for physical objects generally (which is not very well), and can do much better for sorts of events, like deaths and meetings, just as we can for sorts of physical objects, like tables and people. But all this must wait.[28] Meanwhile the situation seems to me to be this: there is a lot of language we can make systematic sense of if we suppose events exist, and we know no promising alternative. The presumption lies with events.

H. *Reply to Cargile*. I suggested that sentences about events and actions be construed as requiring an ontology of particular, unrepeatable, dated events. For example, I argued that a sentence like 'Lucifer fell' has the logical form of an existential quantification of an open sentence true of falls of Lucifer, the open sentence in turn consisting of a two-place predicate true of ordered pairs of things and their falls, and the predicate places filled with a proper name ('Lucifer') and a free variable (bound by the quantifier). I did not explain in detail what I meant by logical form, though I did devote some paragraphs to the subject. I suppose I thought the problems set, the examples and counterexamples offered, the arguments given and the answers entertained would, taken with the tradition and my hints, make the idea clear enough. I was wrong; and in retrospect I sympathize with my misunderstanders. I will try to do better.

Logical form was invented to contrast with something else that is held to be apparent but mere: the form we are led to assign to sentences by superficial analogy or traditional grammar. What meets the eye or ear in language has the charm, complexity, convenience, and deceit of other conventions of the market place, but underlying it is the solid currency of a plainer, duller structure, without wit but also without pretence. This true coin, the deep structure, need never feature directly in the transactions of real life. As long as we know how to redeem our paper we can enjoy the benefits of credit.

The image may help explain why the distinction between logical form and surface grammar can flourish without anyone ever quite explaining it. But what can we say to someone who wonders

---

[28] See Essay 8.

whether there is really any gold in the vaults? I think the concept of logical form can be clarified and thus defended; but the account I shall offer disclaims some of what is implied by the previous paragraph.

What do we mean when we say that 'Whales are mammals' is a quantified sentence? James Cargile suggests that the sentence is elliptical for 'All whales are mammals' (or 'Some whales are mammals') and once the ellipsis is mended we see that the sentence is quantified. Someone interested in logical form would, of course, go much further: he would maintain that 'All whales are mammals' is a universally quantified conditional whose antecedent is an open sentence true of whales and whose consequent is an open sentence true of mammals. The contrast with surface grammar is striking. The subject-predicate analysis goes by the board, 'all whales' is no longer treated as a unit, and the role of 'whales' and of 'mammals' is seen, or claimed, to be predicative.

What can justify this astonishing theory? Part of the answer—the part with which we are most familiar—is that inference is simplified and mechanized when we rewrite sentences in some standardized notation. If we want to deduce 'Moby Dick is a mammal' from 'All whales are mammals' and 'Moby Dick is a whale', we need to connect the predicate 'is a whale' in some systematic way with a suitable feature of 'All whales are mammals'. The theory of the logical form of this sentence tells us how.

Words for temporal precedence, like 'before' and 'after', provide another example. 'The inflation came after the war' is simple enough, at least if we accept events as entities, but how about 'Earwicker slept before Shem kicked Shaun'? Here 'before' connects expressions with the grammatical form of sentences. How is this 'before' related to the one that stands between descriptions of events? Is it a sentential connective, like 'and'? 'Earwicker slept before Shem kicked Shaun' does entail both 'Earwicker slept' and 'Shem kicked Shaun'. Yet clearly 'before' is not truth-functional, since reversing the order of the sentences does not preserve truth.

The solution proposed by Frege has been widely (if not universally) accepted; it is, as we all know, to think of our sentence as doubly quantified by existential quantifiers, to introduce extra places into the predicates to accommodate variables ranging over times, and to interpret 'before' as a two-place predicate. The result, roughly, is: 'There exist two times, t and u, such that Earwicker slept at *t*, Shem kicked Shaun at *u*, and *t* was before *u*.' This analysis relates the two uses of 'before', and it explains why 'Earwicker slept before Shem kicked Shaun' entails 'Shem kicked Shaun'. It does this, however, only by attributing to 'Shem kicked Shaun' the following form: 'There exists a *t* such that Shem kicked Shaun at *t*.' According to Cargile, not even Russell would have denied that '*x* kicked *y*' is a two-place relation form; but Russell had the same motive Frege did for holding that 'kicked' has the logical form of a three-place predicate.

The logical form that the problem of 'before' prompts us to assign to 'Shem kicked Shaun' and to its parts is the very form I suggested, though my reasons were somewhat different, and the ontology was different. So far as ontology is concerned, the two proposals may to advantage be merged, for we may think of 'before' and 'after' as relating events as easily as times. For most purposes, if not all, times are like lengths—convenient abstractions with which we can dispense in favour of the concreta that have them. A significant bonus from this analysis of sentences of temporal priority is that singular causal sentences can then be naturally related to them. According to Hume, if *x* caused *y*, then *x* preceded *y*. What are the entities these variables range over? Events, to be sure. But if this answer is to be taken seriously, then a sentence like 'Sandy's rocking the boat caused it to sink' must somehow refer to events. It does, if we analyse it along these lines: 'There exist two events, *e* and *f*, such that *e* is a rocking of the boat by Sandy, *f* is a sinking of the boat, and *e* caused *f*.'[29]

Let us suppose, as Cargile seems willing to do, that I am right to this extent: by rewriting or rephrasing certain sentences into sentences that explicitly refer to or quantify over events, we can conveniently represent the entailment relations between the original sentences. The entailments we preanalytically recognize to hold between the original sentences become matters of quantificational logic applied to their rephrasals. And now Cargile asks: how can this project, assuming it to be successfully carried out, justify the claim that the original sentences have the logical form of their rewrites? Why not admit that the rewrites show, in many cases, a different form?

Here we have, I hope, the makings of a reconciliation, for I am

---

[29] I discuss this analysis of singular causal sentences in Essay 7.

happy to admit that much of the interest in logical form comes from an interest in logical geography: to give the logical form of a sentence is to give its logical location in the totality of sentences, to describe it in a way that explicitly determines what sentences it entails and what sentences it is entailed by. The location must be given relative to a specific deductive theory; so logical form itself is relative to a theory. The relatively does not stop here, either, since even given a theory of deduction there may be more than one total scheme for interpreting the sentences we are interested in and that preserves the pattern of entailments. The logical form of a particular sentence is, then, relative both to a theory of deduction and to some prior determinations as to how to render sentences in the language of the theory.

Seen in this light, to call the paraphrase of a sentence into some standard first-order quantificational form *the* logical form of the sentence seems arbitrary indeed. Quantification theory has its celebrated merits, to be sure: it is powerful, simple, consistent, and complete in its way. Not least, there are more or less standard techniques for paraphrasing many sentences of natural languages into quantificational languages, which helps excuse not making the relativity to a theory explicit. Still, the relativity remains.

Since there is no eliminating the relativity of logical form to a background theory, the only way to justify particular claims about logical form is by showing that they fit sentences into a *good* theory, at least a theory better than known alternatives. In calling quantificational form logical form I was assuming, like many others before me, that quantification theory is a good theory. What's so good about it?

Well, we should not sneeze at the virtues mentioned above, its known consistency and completeness (in the sense that all quantificational truths are provable). Cargile takes me to task for criticizing Reichenbach's analysis of sentences about events, which introduces an operator that, when prefixed to a sentence, results in a singular term referring to an event. I give a standard argument to show that on this analysis, if one keeps substitutivity of identity and a simple form of extensionality, all events collapse into one. I concluded, 'This demonstrates, I think, that Reichenbach's analysis is radically defective'. Cargile protests that Reichenbach gets in no trouble if the assumption of extensionality is abandoned; and the assumption is mine, not Reichenbach's. Fair enough; I ought not to have said the analysis was defective, but rather that on a natural assumption there was a calamitous consequence. Without the assumption there is no such consequence; but also no theory. Standard quantification theory plus Reichenbach's theory of event sentences plus substitutivity of identity in the new contexts leads to collapse of all events into one. Reichenbach indirectly commits himself to the principle of substitutivity, and Cargile goes along explicitly. So they are apparently committed to giving up standard quantification theory. Since neither offers a substitute, it is impossible to evaluate the position.[30]

Cargile has another idea, which is not to tamper with quantification theory, but simply to add some extra rules to it. If we give the quantificational form of 'Jones buttered the toast in the bathroom' as 'Buttered$_3$ (Jones, the toast, the bathroom)' and of 'Jones buttered the toast' as 'Buttered$_2$ (Jones, the toast)' then the inference from the first to the second is no longer a matter of quantificational logic; but why not interpret this as showing that quantificational form isn't logical form, and quantificational logic isn't all of logic? Cargile suggests that we might be able to give a purely formal (syntactical) rule that would systematize these inferences. I think Cargile underestimates the difficuties in doing this, particularly if, as I have argued, such an approach forces us to admit predicates with indefinitely large numbers of predicate places. I also think he slights the difference, in his remark that 'the standard symbolism of quantification theory is not good at keeping track of entailments between relational forms in English', between simple axioms (which are all that are needed to keep track of the entailments between relational forms in, say, a theory of measurement) and new rules of inference (or axiom schemata). But harping on the difficulties, unless they can be proven to be impossibilities, is inconclusive. It will be more instructive to assume that we are presented with a satisfactory deductive system that adds to quantification theory rules adequate to implement the entailments between event sentences of the sort under consideration. What could then be said in defence of my analysis?

What can be said comes down to this: it explains more, and it explains better. It explains more in the obvious sense of bringing more data under fewer rules. Given my account of the form of

[30] For a discussion of the difficulties of combining substitutivity of identity and non-extensionality, see Dagfinn Føllesdal, 'Quine on Modality'.

sentences about events and actions, certain entailments are a matter of quantificational logic; an account of the kind Cargile hopes to give requires quantificational logic, and then some. But there is a deeper difference.

We catch sight of the further difference if we ask ourselves why 'Jones buttered the toast in the bathroom' entails 'Jones buttered the toast'. So far, Cargile's only answer is, because 'buttered' and some other verbs (listed or characterized somehow) work that way; and my only answer is, because (given my paraphrases) it follows from the rules of quantification theory. But now suppose we ask, *why* do the rules endorse this inference? Surely it has something to do with the fact that 'buttered' turns up in both sentences? There must be a common conceptual element represented by this repeated syntactic feature; we would have a clue to it, and hence a better understanding of the meaning of the two sentences, if we could say what common role 'buttered' has in the two sentences. But here it is evident that Cargile's rules, if they were formulated, would be no help. These rules treat the fact that the word 'buttered' turns up in both sentences as an accident: the rule would work as well if unrelated words were used for the two- and for the three-place predicates. In the analysis I have proposed, the word 'buttered' is discovered to have a common role in the two sentences: in both cases it is a predicate satisfied by certain ordered triples of agents, things buttered, and events. So now we have the beginnings of a new sort of answer to the question why one of our sentences entails the other: it depends on the fact that the word 'buttered' is playing a certain common role in both sentences. By saying exactly what the role is, and what the roles of the other significant features of the sentences are, we will have a deep explanation of why one sentence entails the other, an explanation that draws on a systematic account of how the meaning of each sentence is a function of its structure.

To exhibit an entailment as a matter of quantificational form is to explain it better because we do not need to take the rules of quantificational logic on faith; we can show that they are *valid*, i.e. truth-preserving, by giving an account of the conditions under which sentences in quantificational form are true. From such an account (a theory of truth satisfying Tarski's criteria) it can be seen that if certain sentences are true, others must be. The rules of quantificational logic are justified when we demonstrate that from truths they can lead only to truths.

Plenty of inferences that some might call logical cannot be shown to be valid in any interesting way by appeal to a theory of truth, for example the inference to '*a* is larger than *c*' from '*a* is larger than *b* and *b* is larger than *c*' or to 'Henry is not a man' from 'Henry is a frog'. Clearly a recursive account of truth can ignore these entailments simply by ignoring the logical features of the particular predicates involved. But if I am right, it may not be possible to give a coherent theory of truth that applies to sentences about events and that does not validate the adverbial inferences we have been discussing.

Let me state in more detail how I think our sample inference can be shown to be valid. On my view, a theory of truth would entail that 'Jones buttered the toast in the bathroom' is true if and only if there exists an event satisfying these two conditions: it is a buttering of the toast by Jones, and it occurred in the bathroom. But if these conditions are satisfied, then there is an event that is a buttering of the toast by Jones, and this is just what must be the case, according to the theory, if 'Jones buttered the toast' is true. I put the matter this way because it seems to me possible that Cargile may agree with what I say, then adding, 'But how does this show that "Jones buttered the toast" is a three-place predicate?' If this *is* his response, our troubles are over, or anyway are merely verbal, for all I *mean* by saying that 'Jones buttered the toast' has the logical form of an existentially quantified sentence, and that 'buttered' is a three-piece predicate, is that a theory of truth meeting Tarski's criteria would entail that this sentence is true if and only if there exists ... etc. By my lights, we have given the logical form of a sentence when we have given the truth-conditions of the sentence in the context of a theory of truth that applies to the language as a whole. Such a theory must identify some finite stock of truth-relevant elements, and explicitly account for the truth-conditions of each sentence by how these elements feature in it; so to give the logical form of a sentence is to describe it as composed of the elements the theory isolates.

These remarks will help, I hope, to put talk of 'paraphrasing' or 'translating' in its place. A theory of truth entails, for each sentence *s* of the object language, a theorem of the form '*s* is true if and only if *p*'. Since the sentence that replaces '*p*' must be true (in the metalanguage) if and only if *s* is true (in the object language), there is a sense in which the sentence that replaces '*p*' may be called a translation of *s*; and if the metalanguage contains the object language, it may be called a paraphrase. (These claims must be modified in important ways in a theory of truth for a natural language.) But it should be emphasized that paraphrasis or translation serves no purpose here except that of giving a systematic account of truth-conditons. There is no further claim to synonymy, nor interest in regimentation or improvement. A theory of truth gives a point to such concepts as meaning, translation, and logical form; it does not depend on them.[31]

It should now be clear that my only reason for 'rendering' or 'paraphrasing' event sentences into quantificational form was as a way of giving the truth-conditions for those sentences within a going theory of truth. We have a clear semantics for first-order quantificational languages, and so if we can see how to paraphrase sentences in a natural language into quantificational form, we see how to extend a theory of truth to those sentences. Since the entailments that depend on quantificational form can be completely formalized, it is an easy test of our success in capturing logical form within a theory of truth to see whether our paraphrases articulate the entailments we independently recognize as due to form.

To give the logical form of a sentence is, then, for me, to describe it in terms that bring it within the scope of a semantic theory that meets clear requirements. Merely providing formal rules of inference, as Cargile suggests, thus fails to touch the question of logical form (except by generalizing some of the data a theory must explain); showing how to put sentences into quantificational form, on the other hand, does place them in the context of a semantic theory. The contrast is stark, for it is the contrast between having a theory, and hence a hypothesis about logical form, and having no theory, and hence no way of making sense of claims about form. But of course this does not show that a theory based on first-order quantificational structure and its semantics is all we need or can have. Many philosophers and logicians who have worked on the problem of event sentences (not to mention modalities, sentences about propositional attitudes, and so on) have come to the conclusion that a richer semantics is required, and can be provided. In Essay 6 I explicitly put to one side several obvious problems that invite appeal to such richer schemes. For various reasons I thought, or hoped, that the problem I isolated could be handled within a

---

[31] These claims and others made here are expanded and defended in my 'Truth and Meaning' and 'True to the Facts'.

fairly austere scheme. But when other problems are also emphasized, it may well be that my simple proposal loses its initial appeal;
at least the theory must be augmented, and perhaps it will have to be
abandoned.

Cargile thinks that instead of suggesting that 'Shem kicked
Shaun' has a logical form that is made more explicit by '($\exists x$)
(Kicked (Shem, Shaun, x))' I ought (at most) to have said that the
two sentences are logically equivalent (but have different logical
forms). He makes an analogous point in defending Reichenbach
against my strictures. I want to explain why I resist this adjustment.

Of course it can happen that two sentences are logically equivalent, yet have different logical forms; for example a sentence with
the logical form of a conjunction is logically equivalent to the
conjunction which takes the conjuncts in reverse order. Here we
assume the theory gives the truth-conditions of each sentence, and it
will be possible to prove that one sentence is true if and only if the
other is. But the theory doesn't independently supply truth-
conditions for 'Shem kicked Shaun' and its canonical counterpart;
rather the latter gives (or, in the present discussion, is used to
suggest) the truth conditions of the former. If the theory were
turned on itself, as well it might be, the sentence used to give the
truth-conditions of '($\exists x$) (Kicked (Shem, Shaun, x))' would have
the same form as this sentence; under some natural conditions, it
would be this sentence. So there is no way within the theory of
assigning different logical forms to 'Shem kicked Shaun' and its
explicitly quantificational stand-in. Outside a theory, the notion of
logical form has no clear application, as we have noted. That the two
sentences have very different syntactical structures is evident; that
is why the claim that the logical form is the same is interesting and, if
correct, revealing.

Suppose that a rule of inference is added to our logic, making
each of the two sentences deducible from the other. (Reichenbach
may have had this in mind: see *Elements of Symbolic Logic*, sect.
48.) Will this then make it possible to hold that the sentences have
different logical forms? The answer is as before: rules of inference
that are not backed by a semantic theory are irrelevant to logical
form.

I would like to mention very briefly another point on which
Cargile may have misunderstood me. He says that, 'The idea that
philosophical "analysis" consists in this revealing of logical form is a

popular one . . .' and he may think I share this notion. I don't, and I said that I didn't on the first page of the article he discusses. Even if philosophical analysis were concerned only with language (which I do not believe), revealing logical form would be only part of the enterprise. To know the logical form of a sentence is to know, in the context of a comprehensive theory, the semantic roles of the significant features of the sentence. Aside from the logical constants, this knowledge leaves us ignorant of the relations between predicates, and of their logical properties. To know the logical form of 'The rain caused the flood' is to know whether 'caused' is a sentential connective or a two-place predicate (or something else), but it hardly begins to be knowledge of an analysis of the concept of causality (or the word 'caused'). Or perhaps it is the beginning; but that is all.

On the score of ontology, too, the study of logical form can carry us only a certain distance. If I am right, we cannot give a satisfactory account of the semantics of certain sentences without recognizing that if any of those sentences are true, there must exist such things as events and actions. Given this much, a study of event sentences will show a great deal about what we assume to be true concerning events. But deep metaphysical problems will remain as to the nature of these entities, their mode of individuation, their relations to other categories. Perhaps we will find a way of reducing events to entities of other kinds, for example, sets of points in space-time, or ordered n-tuples of times, physical objects, and classes of ordered n-tuples of such. Successful reductions along these lines may, in an honoured tradition, be advertised as showing that there are no such things as events. As long as the quantifiers and variables remain in the same places, however, the analysis of logical form will stick.

I. *Reply to Hedman*. If we are committed to events, we are committed to making sense of identity-sentences, like '$a = b$', where the terms flanking the identity sign refer to events. I think in fact we use sentences in this form constantly: 'The third round of the fight was (identical with) the one in which he took a dive', 'Our worst accident was (identical with) the one where we hit four other cars', 'Falling off the tower was (identical with) the cause of his death'. The problem of individuation for events is the problem of giving criteria saying when such sentences are true. Carl Hedman raises a tricky question about these criteria as applied to actions.

In Essay 6 I asserted, as Hedman says, that 'intentional actions are not a class of actions'. I said this to protect my theory against an obvious objection. If 'intentional' modifies actions the way 'in the kitchen' does, then intentional actions *are* a class of actions. Does Oedipus's striking of the rude old man belong in this class or not? Oedipus struck the rude old man intentionally, but he did not strike his father intentionally. But on my theory, these strikings were one, since the rude old man was Oedipus's father. The obvious solution, which I endorsed, is to take 'intentionally' as creating a semantically opaque context in which one would expect substitutivity of identity to seem to fail.

I did not argue for this view in the article Hedman discusses; in the long passage he quotes I say only that it is 'the natural and, I think, correct answer'. In that passage I was surveying a number of topics, such as causality, theory of action, explanations, and the identity theory of mind, where philosophers tend to say things which take for granted an ontology of events and actions. My point was that if they do make this assumption, they ought to come up with a serious theory about how reference to events occurs; my intention was to soften up potential opposition to the analysis which (I argued) is forced on us anyway when we try to give a systematic semantics for natural language.

Elsewhere I have argued for the view that one and the same action may be correctly said to be intentional (when described in one way) and not intentional (when described in another). The position is hardly new with me; it was expounded at length by Anscombe,[32] and has been accepted by many other writers on action. It is harder to avoid taking this position than one might think. I suppose no one wants to deny that if the rude old man was Oedipus's father, then 'Oedipus struck the rude old man' and 'Oedipus struck his father' entail one another. If one accepts, as Hedman apparently does, an ontology of events, one will also presumably want to infer 'The striking of the rude old man by Oedipus occurred at the crossroads' from 'The striking of Oedipus's father by Oedipus occurred at the crossroads' and vice versa. But how can these entailments be shown (by a semantical theory) to be valid without also proving the following to be true: 'The striking of the rude old man by Oedipus was identical with the striking of

[32] G. E. M. Anscombe, *Intention*.

Oedipus's father by Oedipus'? Yet one of these actions was intentional and the other not. I don't say no theory could be contrived to validate the wanted inferences while not endorsing the identity; but we don't now have such a theory.

7 *Causal Relations*

What is the logical form of singular causal statements like: 'The flood caused the famine', 'The stabbing caused Caesar's death', 'The burning of the house caused the roasting of the pig'? This question is more modest than the question how we know such statements are true, and the question whether they can be analysed in terms of, say, constant conjunction. The request for the logical form is modest because it is answered when we have identified the logical or grammatical roles of the words (or other significant stretches) in the sentences under scrutiny. It goes beyond this to define, analyse, or set down axioms governing, particular words or expressions.

I

According to Hume, 'we may define a cause to be an object, followed by another, and where all the objects similar to the first are followed by objects similar to the second'. This definition pretty clearly suggests that causes and effects are entities that can be named or described by singular terms; probably events, since one can follow another. But in the *Treatise*, under 'rules by which to judge of causes and effects', Hume says that 'where several different objects produce the same effect, it must be by means of some quality, which we discover to be common amongst them. For as like effects imply like causes, we must always ascribe the causation to the circumstances, wherein we discover the resemblance.' Here it seems to be the 'quality' or 'circumstances' of an event that is the cause rather than the event itself, for the event itself is the same as others in some respects and different in other respects. The suspicion that it is not events, but something more closely tied to the descriptions of events, that Hume holds to be causes, is fortified by Hume's claim that causal statements are never necessary. For if events were causes, then a true description of some event would be 'the cause of *b*', and, given that such an event exists, it follows logically that the cause of *b* caused *b*.

Mill said that the cause 'is the sum total of the conditions positive and negative taken together ... which being realized, the consequent invariably follows'. Many discussions of causality have concentrated on the question whether Mill was right in insisting that the 'real Cause' must include all the antecedent conditions that jointly were sufficient for the effect, and much ingenuity has been spent on discovering factors, pragmatic or otherwise, that guide and justify our choice of some 'part' of the conditions as the cause. There has been general agreement that the notion of cause may be at least partly characterized in terms of sufficient and (or) necessary conditions.[1] Yet it seems to me we do not understand how such characterizations are to be applied to particular causes.

Take one of Mill's examples: some man, say Smith, dies, and the cause of his death is said to be that his foot slipped in climbing a ladder. Mill would say we have not given the whole cause, since having a foot slip in climbing a ladder is not always followed by death. What we were after, however, was not the cause of death in general but the cause of Smith's death: does it make sense to ask under what conditions Smith's death invariably follows? Mill suggests that part of the cause of Smith's death is 'the circumstance of his weight', perhaps because if Smith had been light as a feather his slip would not have injured him. Mill's explanation of why we don't bother to mention this circumstance is that it is too obvious to bear mention, but it seems to me that if it was Smith's fall that killed him, and Smith weighed twelve stone, then Smith's fall was the fall of a man who weighed twelve stone, whether or not we know it or mention it. How could Smith's actual fall, with Smith weighing, as he did, twelve stone, be any more efficacious in killing him than Smith's actual fall?

The difficulty has nothing to do with Mill's sweeping view of the cause, but attends any attempt of this kind to treat particular causes

[1] For an example, with references to many others, see J. L. Mackie, 'Causes and Conditions'.

as necessary or sufficient conditions. Thus Mackie asks, 'What is the exact force of [the statement of some experts] that this short-circuit caused this fire?' And he answers, 'Clearly the experts are not saying that the short-circuit was a necessary condition for this house's catching fire at this time; they know perfectly well that a short-circuit somewhere else, or the overturning of a lighted oil stove ... might, if it had occurred, have set the house on fire.'[2] Suppose the experts know what they are said to; how does this bear on the question whether the short circuit was a necessary condition of this particular fire? For a short circuit elsewhere could not have caused *this* fire, nor could the overturning of a lighted oil stove.

To talk of particular events as conditions is bewildering, but perhaps causes aren't events (like the short circuit, or Smith's fall from the ladder), but correspond rather to sentences (perhaps like the fact that this short circuit occurred, or the fact that Smith fell from the ladder). For sentences can express conditions of truth for other sentences.

If causes correspond to sentences rather than singular terms, the logical form of a sentence like:

(1) The short circuit caused the fire.

would be given more accurately by:

(2) *The fact that* there was a short circuit *caused it to be the case that* there was a fire.

In (2) the italicized words constitute a sentential connective like 'and' or 'if ... then ....' This approach no doubt receives support from the idea that causal laws are universal conditionals, and singular causal statements ought to be instances of them. Yet the idea is not easily implemented. Suppose, first, that a causal law is (as it is usually said Hume taught) nothing but a universally quantified material conditional. If (2) is an instance of such, the italicized words have just the meaning of the material conditional, 'If there was a short circuit, then there was a fire'. No doubt (2) entails this, but not conversely, since (2) entails something stronger, namely the conjunction 'There was a short circuit *and* there was a fire'. We might try treating (2) as the conjunction of the appropriate law and 'There was a short circuit and there was a fire'—indeed this seems a possible interpretation of Hume's definition of cause quoted

[2] Ibid., 245.

above—but then (2) would no longer be an instance of the law. And aside from the inherent implausibility of this suggestion as giving the logical form of (2) (in contrast, say, to giving the grounds on which it might be asserted) there is also the oddity that an inference from the fact that there was a short circuit and there was a fire, and the law, to (2) would turn out to be no more than a conjoining of the premises.

Suppose, then that there is a non-truth-functional causal connective, as has been proposed by many.[3] In line with the concept of a cause as a condition, the causal connective is conceived as a conditional, though stronger than the truth-functional conditional. Thus Arthur Pap writes, 'The distinctive property of causal implication as compared with material implication is just that the falsity of the antecedent is no ground for inferring the truth of the causal implication'. If the connective Pap had in mind were that of (2), this remark would be strange, for it is a property of the connective in (2) that the falsity of either the 'antecedent' or the 'consequent' is a ground for inferring the falsity of (2). That treating the causal connective as a kind of conditional unsuits it for the work of (1) or (2) is perhaps even more evident from Burks's remark that 'p is causally sufficient for q is logically equivalent to $\sim q$ is causally sufficient for $\sim p$'. Indeed, this shows not only that Burks's connective is not that of (2), but also that it is not the subjunctive causal connective 'would cause'. My tickling Jones would cause him to laugh, but his not laughing would not cause it to be the case that I didn't tickle him.

These considerations show that the connective of (2), and hence by hypothesis of (1), cannot, as is often assumed, be a conditional of any sort, but they do not show that (2) does not give the logical form of singular causal statements. To show this needs a stronger argument, and I think there is one, as follows.

It is obvious that the connective in (2) is not truth-functional, since (2) may change from true to false if the contained sentences are switched. Nevertheless, substitution of singular terms for others with the same extension in sentences like (1) and (2) does not touch their truth-value. If Smith's death was caused by the fall from the ladder and Smith was the first man to land on the moon, then the fall from the ladder was the cause of the death of the first man to land on the moon. And if the fact that there was a fire in Jones's house

[3] For example by: J. L. Mackie, op. cit., 254; Arthur Burks, 'The Logic of Causal Propositions', 369; and Arthur Pap, 'Disposition Concepts and Extensional Logic', 212.

caused it to be the case that the pig was roasted, and Jones's house is the oldest building on Elm Street, then the fact that there was a fire in the oldest building on Elm Street caused it to be the case that the pig was roasted. We must accept the principle of extensional substitution, then. Surely also we cannot change the truth value of the likes of (2) by substituting logically equivalent sentences for sentences in it. Thus (2) retains its truth if for 'there was a fire' we substitute the logically equivalent '$\hat{x}$ ($x = x$ & there was a fire) = $\hat{x}$ ($x = x$)'; retains it still if for the left side of this identity we write the coextensive singular term '$\hat{x}$ ($x = x$ & Nero fiddled)'; and still retains it if we replace '$\hat{x}$ ($x = x$ & Nero fiddled) = $\hat{x}$ ($x = x$)' by the logically equivalent 'Nero fiddled.' Since the only aspect of 'there was a fire' and 'Nero fiddled' that matters to this chain of reasoning is the fact of their material equivalence, it appears that our assumed principles have led to the conclusion that the main connective of (2) is, contrary to what we supposed, truth-functional.[4]

Having already seen that the connective of (2) cannot be truth-functional, it is tempting to try to escape the dilemma by tampering with the principles of substitution that led to it. But there is another, and, I think, wholly preferable way out: we may reject the hypothesis that (2) gives the logical form of (1), and with it the ideas that the 'caused' of (1) is a more or less concealed sentential connective, and that causes are fully expressed only by sentences.

II

Consider these six sentences:

(3) *It is a fact that* Jack fell down.
(4) Jack fell down *and* Jack broke his crown.
(5) Jack fell down *before* Jack broke his crown.
(6) Jack fell down, *which caused it to be the case that* Jack broke his crown.
(7) *Jones forgot the fact that* Jack fell down.
(8) *That* Jack fell down *explains the fact that* Jack broke his crown.

[4] This argument is closely related to one spelled out by Dagfinn Føllesdal (in 'Quantification into Causal Contexts') to show that unrestricted quantification into causal contexts leads to difficulties. His argument is in turn a direct adaptation of Quine's (*Word and Object*, 197–8) to show that (logical) modal distinctions collapse under certain natural assumptions. The argument derives from Frege.

Substitution of equivalent sentences for, or substitution of coextensive singular terms or predicates in, the contained sentences, will not alter the truth-value of (3) or (4): here extensionality reigns. In (7) and (8), intentionality reigns, in that similar substitution in or for the contained sentences is not guaranteed to save truth. (5) and (6) seem to fall in between; for in them substitution of coextensive singular terms preserves truth, whereas substitution of equivalent sentences does not. However this last is, as we just saw with respect to (2), and hence also (6), untenable middle ground.

Our recent argument would apply equally against taking the 'before' of (5) as the sentential connective it appears to be. And of course we don't interpret 'before' as a sentential connective, but rather as an ordinary two-place relation true of ordered pairs of times; this is made to work by introducing an extra place into the predicates ('x fell down' becoming 'x fell down at t') and an ontology of times to suit. The logical form of (5) is made perspicuous, then, by:

(5) There exists times t and t' such that Jack fell down at t, Jack broke his crown at t', and t preceded t'.

This standard way of dealing with (5) seems to me essentially correct and I propose to apply the same strategy to (6), which then comes out:

(6') There exist events e and e' such that e is a falling down of Jack, e' is a breaking of his crown by Jack, and e caused e'.

Once events are on hand, an obvious economy suggests itself: (5) may as well be construed as about events rather than times. With this, the canonical version of (5) becomes just (6'), with 'preceded' replacing 'caused'. Indeed, it would be difficult to make sense of the claim that causes precede, or at least do not follow, their effects if (5) and (6) did not thus have parallel structures. We will still want to be able to say when an event occurred, but with events this requires an ontology of pure numbers only. So, 'Jack fell down at 3 p.m.' says that there is an event e that is a falling down of Jack, and the time of e measured in hours after noon, is three; more briefly, $(\exists e) (F (\text{Jack}, e) \ \& \ t(e) = 3)$.

On the present plan, (6) means some fall of Jack's caused some breaking of Jack's crown; so (6) is not false if Jack fell more than once, broke his crown more than once, or had a crown-breaking fall more than once. Nor, if such repetitions turned out to be the case, would we have grounds for saying that (6) referred to one rather than another of the fracturings. The same does not go for 'The short circuit caused the fire' or 'The flood caused the famine' or 'Jack's fall caused the breaking of Jack's crown'; here singularity is imputed. ('Jack's fall', like 'the day after tomorrow', is no less a singular term because it may refer to different entities on different occasions.) To do justice to 'Jack's fall caused the breaking of Jack's crown' what we need is something like 'The one and only falling down of Jack caused the one and only breaking of his crown by Jack'; in some symbols of the trade, '$(\imath e) F$ (Jack, e) caused $(\imath e) B$ (Jack's crown, e)'.

Evidently (1) and (2) do not have the same logical form. If we think in terms of standard notations for first-order languages, it is (1) that more or less wears its form on its face; (2), like many existentially quantified sentences, does not (witness 'Somebody loves somebody'). The relation between (1) and (2) remains obvious and close: (1) entails (2), but not conversely.[5]

III

The salient point that emerges so far is that we must distinguish firmly between causes and the features we hit on for describing them, and hence between the question whether a statement says truly that one event caused another and the further question whether the events are characterized in such a way that we can deduce, or otherwise infer, from laws or other causal lore, that the relation was causal. 'The cause of this match's lighting is that it was struck.—Yes, but that was only *part* of the cause; it had to be a dry match, there had to be adequate oxygen in the atmosphere, it had to be struck hard enough, etc.' We ought now to appreciate that the 'Yes, but' comment does not have the force we thought. It cannot be that the striking of this match was only part of the cause, for this match was in fact dry, in adequate oxygen, and the striking was hard

[5] A familiar device I use for testing hypotheses about logical grammar is translation into standard quantificational form; since the semantics of such languages is transparent, translation into them is a way of providing a semantic theory (a theory of the logical form) for what is translated. In this employment, canonical notation is not to be conceived as an improvement on the vernacular, but as a comment on it.

For elaboration and defence of the view of events sketched here, see Essay 6.

enough. What is partial in the sentence, 'The cause of this match's lighting is that it was struck' is the *description* of the cause; as we add to the description of the cause, we may approach the point where we can deduce, from this description and laws, that an effect of the kind described would follow.

If Flora dried herself with a coarse towel, she dried herself with a towel. This is an inference we know how to articulate, and the articulation depends in an obvious way on reflecting in language an ontology that includes such things as towels: if there is a towel that is coarse and was used by Flora in her drying, there is a towel that was used by Flora in her drying. The usual way of doing things does not, however, give similar expression to the similar inference from 'Flora dried herself with a towel on the beach at noon' to 'Flora dried herself with a towel', or for that matter, from the last to 'Flora dried herself'. But if, as I suggest, we render 'Flora dried herself' as about an event, as well as about Flora, these inferences turn out to be quite parallel to the more familiar ones. Thus if there was an event that was a drying by Flora of herself and that was done with a towel, on the beach, at noon, then clearly there was an event that was a drying by Flora of herself—and so on.

The mode of inference carries over directly to causal statements. If it was a drying she gave herself with a coarse towel on the beach at noon that caused those awful splotches to appear on Flora's skin, then it was a drying she gave herself that did it; we may also conclude that it was something that happened on the beach, something that took place at noon, and something that was done with a towel, that caused the tragedy. These little pieces of reasoning seem all to be endorsed by intuition, and it speaks well for the analysis of causal statements in terms of events that on that analysis the arguments are transparently valid.

Mill, we are now in better position to see, was wrong in thinking we have not specified the whole cause of an event when we have not wholly specified it. And there is not, as Mill and others have maintained, anything elliptical in the claim that a certain man's death was caused by his eating a particular dish, even though death resulted only because the man had a particular bodily constitution, a particular state of present health, and so on. On the other hand Mill was, I think, quite right in saying that 'there certainly is, among the circumstances that took place, some combination or other on which death is invariably consequent ... the whole of which circumstances perhaps constituted in this particular case the conditions of the phenomenon . . .'.[6] Mill's critics are no doubt justified in contending that we may correctly give the cause without saying enough about it to demonstrate that it was sufficient; but they share Mill's confusion if they think every deletion from the description of an event represents something deleted from the event described.

The relation between a singular causal statement like 'The short circuit caused the fire' and necessary and sufficient conditions seems, in brief, to be this. The fuller we make the description of the cause, the better our chances of demonstrating that it was sufficient (as described) to produce the effect, and the worse our chances of demonstrating that it was necessary; the fuller we make the description of the effect, the better our chances of demonstrating that the cause (as described) was necessary, and the worse our chances of demonstrating that it was sufficient. The symmetry of these remarks strongly suggests that in whatever sense causes are correctly said to be (described as) sufficient, they are as correctly said to be necessary. Here is an example. We may suppose there is some predicate '$P(x,y,e)$' true of Brutus, Caesar, and Brutus's stabbing of Caesar and such that any stab (by anyone) that is P is followed by the death of the stabbed. And let us suppose further that this law meets Mill's requirements of being *unconditional*—it supports counterfactuals of the form, 'If Cleopatra had received a stab that was P, she would have died.' Now we can prove (assuming a man dies only once) that Brutus's stab was sufficient for Caesar's death. Yet it was not the cause of Caesar's death, for Caesar's death was the death of a man with more wounds than Brutus inflicted, and such a death could not have been caused by an event that was P ('P' was chosen to apply only to stabbings administered by a single hand). The trouble here is not that the description of the cause is partial, but that the event described was literally (spatio-temporally) only part of the cause.

Can we then analyse 'a caused b' as meaning that a and b may be described in such a way that the existence of each could be demonstrated, in the light of causal laws, to be a necessary and sufficient condition of the existence of the other? One objection, foreshadowed in previous discussion, is that the analysandum does, but the analysans does not, entail the existence of a and b. Suppose we add, in remedy, the condition that either a or b, as described, exists.

[6] *A System of Logic*, Book III, Chap. V, sect. 3.

Then on the proposed analysis one can show that the causal relation holds between any two events. To apply the point in the direction of sufficiency, imagine some description '$(\imath x)\ Fx$' under which the existence of an event a may be shown sufficient for the existence of b. Then the existence of an arbitrary event c may equally be shown sufficient for the existence of b: just take as the description of c the following: '$(\imath y)\ (y = c\ \&\ (\exists !x)\ Fx)$'.[7] It seems unlikely that any simple and natural restrictions on the form of allowable descriptions would meet this difficulty, but since I have abjured the analysis of the causal relation, I shall not pursue the matter here.

There remains a legitimate question concerning the relation between causal laws and singular causal statements that may be raised independently. Setting aside the abbreviations successful analysis might authorize, what form are causal laws apt to have if from them and a premise to the effect that an event of a certain (acceptable) description exists, we are to infer a singular causal statement saying that the event caused, or was caused by, another? A possibility I find attractive is that a full-fledged causal law has the form of a conjunction:

$$(L)\begin{cases} (S) & (e)\ (n)\ ((Fe\ \&\ t(e) = n) \rightarrow \\ & \qquad (\exists !f)\ (Gf\ \&\ t(f) = n + \varepsilon\ \&\ C(e,f)))\ and \\ (N) & (e)\ (n)\ ((Ge\ \&\ t(e) = n + \varepsilon) \rightarrow \\ & \qquad (\exists !f)\ (Ff\ \&\ t(f) = n\ \&\ C(f,\ e))) \end{cases}$$

Here the variables 'e' and 'f' range over events, 'n' ranges over numbers, F and G are properties of events, '$C(e,f)$' is read 'e causes f', and 't' is a function that assigns a number to an event to mark the time the event occurs. Now, given the premise:

$$(P) \qquad\qquad (\exists !e)\ (Fe\ \&\ t(e) = 3)$$

we may infer:

$$(C)\ (\imath e)\ (Fe\ \&\ t(e) = 3)\ \text{caused}\ (\imath e)\ (Ge\ \&\ t(e) = 3 + \varepsilon)$$

It is worth remarking that part (N) of (L) is as necessary to the proof of (C) from (P) as is to the proof of (C) from the premise '$(\exists !e)\ (Ge\ \&\ t(e) = 3 + \varepsilon)$'. This is perhaps more reason for holding that causes are, in the sense discussed above, necessary as well as sufficient conditions.

---

[7] Here I am indebted to Professor Carl Hempel, and in the next sentence to John Wallace.

Explaining 'why an event occurred', on this account of laws, may take an instructively large number of forms, even if we limit explanation to the resources of deduction. Suppose, for example, we want to explain the fact that there was a fire in the house at 3.01 p.m. Armed with appropriate premises in the form of (P) and (L), we may deduce: that there was a fire in the house at 3.01 p.m.; that it was caused by a short circuit at 3.00 p.m.; that there was only one fire in the house at 3.01 p.m.; that this fire was caused by the one and only short circuit that occurred at 3.00 p.m. Some of these explanations fall short of using all that is given by the premises; and this is lucky, since we often know less. Given only (S) and (P), for example, we cannot prove there was only one fire in the house at 3.01 p.m., though we can prove there was exactly one fire in the house at 3.01 p.m. that was caused by the short circuit. An interesting case is where we know a law in the form of (N), but not the corresponding (S). Then we may show that, given that an event of a particular sort occurred, there must have been a cause answering to a certain description, but, given the same description of the cause, we could not have predicted the effect. An example might be where the effect is getting pregnant.

If we explain why it is that a particular event occurred by deducing a statement that there is such an event (under a particular description) from a premise known to be true, then a simple way of explaining an event, for example the fire in the house at 3.01 p.m., consists in producing a statement of the form of (C); and this explanation makes no use of laws. The explanation will be greatly enhanced by whatever we can say in favour of the truth of (C); needless to say, producing the likes of (L) and (P), if they are known true, clinches the matter. In most cases, however, the request for explanation will describe the event in terms that fall under no full-fledged law. The device to which we will then resort, if we can, is apt to be redescription of the event. For we can explain the occurrence of any event a if we know (L), (P), and the further fact that $a = (ie) (Ge \,\&\, t(e) = 3 + \varepsilon)$. Analogous remarks apply to the redescription of the cause and to cases where all we want to, or can, explain is the fact that there was *an* event of a certain sort.

The great majority of singular causal statements are not backed, we may be sure, by laws in the way (C) is backed by (L). The relation in general is rather this: if 'a caused b' is true, then there are

descriptions of a and b such that the result of substituting them for
'a' and 'b' in 'a' caused 'b' is entailed by true premises of the form of
(L) and (P); and the converse holds if suitable restrictions are put on
the descriptions.[8] If this is correct, it does not follow that we must be
able to dredge up a law if we know a singular causal statement to be
true; all that follows is that we know there must be a covering law.
And very often, I think, our justification for accepting a singular
causal statement is that we have reason to believe an appropriate
causal law exists, though we do not know what it is. Generalizations
like 'If you strike a well-made match hard enough against a properly prepared surface, then, other conditions being favourable, it
will light' owe their importance not to the fact that we can hope
eventually to render them untendentious and exceptionless, but
rather to the fact that they summarize much of our evidence for
believing that full-fledged causal laws exist covering events we wish
to explain.[9]

If the story I have told is true, it is possible to reconcile, within
limits, two accounts thought by their champions to be opposed. One
account agrees with Hume and Mill to this extent: it says that a
singular causal statement 'a caused b' entails that there is a law to
the effect that 'all the objects similar to a are followed by objects
similar to b' and that we have reason to believe the singular statement only in so far as we have reason to believe there is such a law.
The second account (persuasively argued by C. J. Ducasse[10] maintains that singular causal statements entail no law and that we can
know them to be true without knowing any relevant law. Both of
these accounts are entailed, I think, by the account I have given, and
they are consistent (I therefore hope) with each other. The reconciliation depends, of course, on the distinction between knowing
there is a law 'covering' two events and knowing what the law is: in
my view, Ducasse is right that singular causal statements entail no
law; Hume is right that they entail there is a law.

[8] Clearly this account cannot be taken as a definition of the causal relation. Not
only is there the inherently vague quantification over expressions (of what language?), but there is also the problem of spelling out the 'suitable restrictions'.

[9] The thought in these paragraphs, like much more that appears here, was first
adumbrated in Essay 1. This conception of causality was subsequently discussed and,
with various modifications, employed by Samuel Gorovitz, 'Causal Judgments and
Causal Explanations', and by Bernard Berofsky, 'Causality and General Laws'.

[10] See his 'Critique of Hume's Conception of Causality', *Causation and the Types
of Necessity*, and *Nature, Mind, and Death*, Part 11.

IV

Much of what philosophers have said of causes and causal relations is intelligible only on the assumption (often enough explicit) that causes are individual events, and causal relations hold between events. Yet, through failure to connect this basic *aperçu* with the grammar of singular causal judgements, these same philosophers have found themselves pressed, especially when trying to put causal statements into quantificational form, into trying to express the relation of cause to effect by a sentential connective. Hence the popularity of the utterly misleading question: can causal relations be expressed by the purely extensional material conditional, or is some stronger (non-Humean) connection involved? The question is misleading because it confuses two separate matters: the logical form of causal statements and the analysis of causality. So far as form is concerned, the issue of nonextensionality does not arise, since the relation of causality between events can be expressed (no matter how 'strong' or 'weak' it is) by an ordinary two-place predicate in an ordinary, extensional first order language. These plain resources will perhaps be outrun by an adequate account of the form of causal laws, subjunctives, and counterfactual conditionals, to which most attempts to analyse the causal relation turn. But this is, I have urged, another question.

This is not to say there are no causal idioms that directly raise the issue of apparently non-truth-functional connectives. On the contrary, a host of statement forms, many of them strikingly similar, at least at first view, to those we have considered, challenge the account just given. Here are samples: 'The failure of the sprinkling system caused the fire', 'The slowness with which the controls were applied caused the rapidity with which the inflation developed', 'The collapse was caused, not by the fact that the bolt gave way, but by the fact that it gave way so suddenly and unexpectedly', 'The fact that the dam did not hold caused the flood.' Some of these sentences may yield to the methods I have prescribed, especially if failures are counted among events, but others remain recalcitrant. What we must say in such cases is that in addition to, or in place of, giving what Mill calls the 'producing cause', such sentences tell, or suggest, a causal story. They are, in other words, rudimentary causal explanations. Explanations typically relate statements, not events. I suggest therefore that the 'caused' of the sample sentences in this paragraph is not the 'caused' of straightforward singular causal statements, but is best expressed by the words 'causally explains'.[11]

A final remark. It is often said that events can be explained and predicted only in so far as they have repeatable characteristics, but not in so far as they are particulars. No doubt there is a clear and trivial sense in which this is true, but we ought not to lose sight of the less obvious point that there is an important difference between explaining the fact that there was *an* explosion in the broom closet and explaining the occurrence of *the* explosion in the broom closet. Explanation of the second sort touches the particular event as closely as language can ever touch any particular. Of course this claim is persuasive only if there are such things as events to which singular terms may refer. But the assumption, ontological and metaphysical, that there are events, is one without which we cannot make sense of much of our most common talk; or so, at any rate, I have been arguing. I do not know any better, or further, way of showing what there is.

---

[11] Zeno Vendler has ingeniously marshalled the linguistic evidence for a deep distinction, in our use of 'cause', 'effect', and related words, between occurrences of verb-nominalizations that are fact-like or propositional, and occurrences that are event-like. (See Zeno Vendler, 'Effects, Results and Consequences'.) Vendler concludes that the 'caused' of 'John's action caused the disturbance' is always flanked by expressions used in the propositional or fact-like sense; 'was an effect of' or 'was due to' in 'The shaking of the earth was an effect of (was due to) the explosion' is flanked by expressions in the event-like sense. My distinction between essentially sentential expressions and the expressions that refer to events is much the same as Vendler's and owes much to him, though I have used more traditional semantic tools and have interpreted the evidence differently.

My suggestion that 'caused' is sometimes a relation, sometimes a connective, with corresponding changes in the interpretation of the expressions flanking it, has much in common with the thesis of J. M. Shorter's 'Causality, and a Method of Analysis'.

8 *The Individuation of Events*

When are events identical, when distinct? What criteria are there for deciding one way or the other in particular cases?

There is a familiar embarrassment in asking identity questions of this sort that comes out clearly if we rephrase the question slightly: when are two events identical? Or, when is one event identical with another? It seems only one answer is possible: no *two* events are identical, no event is ever identical with *another*. It is hopeless to try to improve matters by asking instead, when is an event identical with itself? For again, only one answer is possible: always.

The difficulty obviously has nothing special to do with events, it arises in relation to all identity questions. The only move I know for circumventing this conundrum is to substitute for questions about identities questions about sentences about identities. Then instead of asking when events are identical, we may ask when sentences of the form '$a = b$' are true, where we suppose 'a' and 'b' supplanted by singular terms referring to events.

We have no sooner to restate our problem in this standard way, however, than to realize something scandalous about events. Events, even in the best philosophical circles, lead a double life. On the one hand, we talk confidently of sentences that 'describe' or 'refer to' events, and of cases where two sentences refer to the same event; we have grown used to speaking of actions (presumably a species of event) 'under a description'. We characterize causal laws as asserting that every event of one sort is followed by an event of another sort, and it is said that explanation in history and science is often of particular events, though perhaps only as those events are described in one way rather than another. But—and this is the other hand—when we turn to the sentences, formalized in standard ways or in our native dialect, that are so familiarly interpreted as describing or referring to events, or as making universal claims about events, we generally find nothing commonly counted as singular terms that could be taken to refer to events. We are told, for example, that on occasion 'He raised his arm' and 'He signalled' describe the same action; yet where are the singular terms in these sentences that could do the describing? 'Whenever a piece of metal is heated it expands' is normally taken as quantifying over physical objects and perhaps times; how could we analyse it so as to justify the claim that it literally speaks of events?

Quine has quipped: 'No entity without identity' in support of the Fregean thesis that we ought not to countenance entities unless we are prepared to make sense of sentences affirming and denying identity of such entities. But then more obvious still is the motto: 'No identity without an entity', and its linguistic counterpart: 'No statements of identity without singular terms'.

Our problem was to determine when sentences formed by flanking an identity sign with singular terms referring to events are true; at this point the problem seems to invite the response that there are no such sentences because there are no such singular terms. But of course this is too strong; there are singular terms that apparently name events: 'Sally's third birthday party', 'the eruption of Vesuvius in A.D. 1906', 'my eating breakfast this morning', 'the first performance of *Lulu* in Chicago'. Still, the existence of these singular terms is of uncertain relevance until we can firmly connect such singular terms with sentences like 'Vesuvius erupted in A.D. 1906' or 'I ate breakfast this morning', for most of our interest in identity sentences about events depends upon the assumption that the singular terms that appear in them refer to entities that are needed for the analysis of more ordinary sentences. If the only pressure for adopting an ontology of events comes from such phrases as 'Sally's third birthday party', we would probably do better to try and paraphrase these away in context than meddle with the logical form of sentences like 'Brutus killed Caesar' or 'Bread nourishes' so as to show singular terms referring to events or variables ranging over them.

Are there good reasons for taking events seriously as entities? There are indeed. First, it is hard to imagine a satisfactory theory of action if we cannot talk literally of the same action under different descriptions. Jones managed to apologize by saying 'I apologize';

but only because, under the circumstances, saying 'I apologize' *was* apologizing. Cedric intentionally burned the scrap of paper; this serves to excuse his burning a valuable document only because he did not know the scrap was the document and because his burning the scrap was (identical with) his burning the document. Explanation, as already hinted, also seems to call for events. Last week there was a catastrophe in the village. In the course of explaining why it happened, we need to redescribe it, perhaps as an avalanche. There are rough statistical laws about avalanches: avalanches tend to occur when a heavy snow falls after a period of melting and freezing, so that the new snow does not bind to the old. But we could go further in explaining this avalanche—why it came just when it did, why it covered the area it did, and so forth—if we described it in still a different and more precise vocabulary. And when we mention, in one way or another, the cause of the avalanche, we apparently claim that though we may not know such a description or such a law, there must *be* descriptions of cause and avalanche such that those descriptions instantiate a true causal law. All this talk of descriptions and redescriptions makes sense, it would seem, only on the assumption that there are *bona fide* entities to be described and redescribed. A further need for events springs from the fact that the most perspicuous forms of the identity theory of mind require that we identify mental events with certain physiological events; if such theories or their denials are intelligible, events must be individuals. And for such theories to be interesting, there must be ways of telling when statements of event-identity are true.[1]

The reasons just canvassed for accepting an explicit ontology of events rest upon the assumption that one or another currently accepted or debated philosophical position or doctrine is intelligible when taken at face value; so it remains possible to resist the conclusion by rejecting the relevant doctrines as unintelligible, or by attempting to reinterpret them without appeal to events. The prospects for successful resistance are, in my opinion, dim: I do not believe we can give a cogent account of action, of explanation, of causality, or of the relation between the mental and the physical, unless we accept events as individuals. Each of these claims needs detailed defence.[2]

[1] This point is well stated by Jaegwon Kim, 'On the Psycho-Physical Identity Theory'.
[2] See Essays 1, 6, 7, and 11.

There remains, however, a more direct consideration (of which the others are symptoms) in favour of an ontology of events, which is that without events it does not seem possible to give a natural and acceptable account of the logical form of certain sentences of the most common sorts; it does not seem possible, that is, to show how the meanings of such sentences depend upon their composition. The situation may be sketched as follows. It is clear that the sentence 'Sebastian strolled through the streets of Bologna at 2 a.m.' entails 'Sebastian strolled through the streets of Bologna', and does so by virtue of its logical form. This requires, it would seem, that the patent syntactical fact that the entailed sentence is contained in the entailing sentence be reflected in the logical form we assign to each sentence. Yet the usual way of formalizing these sentences does not show any such feature: it directs us to consider the first sentence as containing an irreducibly three-place predicate 'x strolled through y at t' while the second contains the unrelated predicate 'x strolled through y'. It is sometimes proposed that we can mend matters by treating 'Sebastian strolled through the streets of Bologna' as elliptical for 'There exists a time t such that Sebastian strolled through the streets of Bologna at t'. This suggestion contains the seed of a general solution, however, only if we can form a clear idea of how many places predicates of action or change involve. But it is unlikely that we can do this since there appear to be ways of adding indefinitely to the number of places that would be required. Consider, for example, 'The shark devoured Danny by chewing up his left foot, then his left ankle, then his left knee, then ...', or, 'The fall of the first domino caused the fall of the last by causing the fall of the second, which caused the fall of the third, which caused. ...'.[3]

Ingenuity may conceive more than one way of coping with these and associated puzzles, but it is impressive how well everything comes out if we accept the obvious idea that there are things like falls, devourings, and strolls for sentences such as these to be about. In short, I propose to legitimize our intuition that events are true particulars by recognizing explicit reference to them, or quantification over them, in much of our ordinary talk. Take as an example, 'Sebastian strolled': this may be construed along lines suggested by 'Sebastian took a stroll.' 'There is an x such that x is a stroll and

---

[3] The difficulty discussed here is raised by Anthony Kenny in *Action, Emotion and Will*, Ch. VII. In Essay 6 I devote more space to these matters and to the solution about to be outlined.

Sebastian took *x*' is more ornate than necessary, since there is nothing an agent can do with a stroll except take it; thus we may capture all there is with 'There is an *x* such that Sebastian strolled *x*.'

In this way we provide each verb of action or change with an event-place; we may say of such verbs that they take an *event-object*. Adverbial modification is thus seen to be logically on a par with adjectival modification: what adverbial clauses modify is not verbs, but the events that certain verbs introduce. 'Sebastian strolled through the streets of Bologna at 2 a.m.' then has this form: 'There is an event *x* such that Sebastian strolled *x*, *x* took place in the streets of Bologna, and *x* was going on at 2 a.m.' Clearly, the entailments that worried us before go through directly on this analysis.

We recognize that there is no singular term referring to a mosquito in 'There is a mosquito in here' when we realize that the truth of this sentence is not impugned if there are two mosquitos in the room. It would not be appropriate if, noticing that there are two mosquitos in the room, I were to ask the person who says, 'There is a mosquito in the room', 'Which one are you referring to?' On the present analysis, ordinary sentences about events, like 'Doris capsized the canoe yesterday', are related to particular events in just the same way that 'There is a mosquito in here' is related to particular mosquitos. It is no less true that Doris capsized the canoe yesterday if she capsized it a dozen times than if she capsized it once; nor, if she capsized it a dozen times, does it make sense to ask, 'Which time are you referring to?' as if this were needed to *clarify* 'Doris capsized the canoe yesterday'. We learned some time ago, and it is a very important lesson, that phrases like 'a mosquito' are not singular terms, and hence do not refer as names or descriptions do. The temptation to treat a sentence like 'Doris capsized the canoe yesterday' as if it contained a singular term referring to an action comes from other sources, but we should be equally steadfast in resisting it.

Some actions are difficult or unusual to perform more than once in a short or specified time, and this may provide a specious reason in some cases for holding that action sentences refer to unique actions. Thus with 'Jones got married last Saturday', 'Doris wrote a cheque at noon', 'Mary kissed an admirer at the stroke of midnight'. It is merely illegal to get married twice on the same day, merely unusual to write cheques simultaneously, and merely good fortune

168   *Event and Cause*

to get to kiss two admirers at once. Similarly, if I say, 'There is an elephant in the bathtub', you are no doubt justified in supposing that one elephant at most is in the bathtub, but you are confused if you think my sentence contains a singular term that refers to a particular elephant if any. A special case arises when we characterize actions in ways that logically entail that at most one action so characterized exists: perhaps you can break a certain piece of news to a particular audience only once; a man can assassinate his enemy only once; a woman can lose her virtue only once. 'Brutus killed Caesar' is then arguably equivalent to 'Brutus killed Caesar exactly once' which is arguably equivalent (by way of Russell's theory of descriptions) to 'The killing of Caesar by Brutus occurred.' This last certainly does contain a description, in the technical sense, of an action, and so we could say that 'Brutus killed Caesar' refers to or describes the killing of Caesar by Brutus in that it is logically equivalent to a sentence that overtly refers to or describes the killing of Caesar by Brutus. By parity of reasoning we should, of course, maintain that 'There exists a prime between 20 and 28' refers to the number 23. There is a good reason against taking this line, however, which is that on this view someone could be uniquely referring without knowing he was using words that imputed singularity.

Confusion over the relation between ordinary sentences about actions, and particular actions, has led some philosophers to suppose or to suggest that these sentences are about *generic* actions, or *kinds* of actions. Von Wright, for example, says that 'Brutus killed Caesar' is about a particular action, while 'Brutus kissed Caesar' is about a generic action.[4] It is true that we can paraphrase 'Brutus kissed Caesar' as 'There is at least one event belonging to the genus, a kissing of Caesar by Brutus'; but we can equally well paraphrase 'Brutus killed Caesar' as 'There is at least one event belonging to the genus, a killing of Caesar by Brutus.' In neither case does the sentence refer to a generic action. Analogous remarks apply to the idea that 'Lying is wrong' is about a kind of action. 'Lying is wrong' may be rendered, 'For all x if x is a lie then x is wrong' or even, 'The class of lies is included in the class of wrong actions', but neither of these says that a kind of action is wrong, but rather that each action of a kind is wrong.

Failure to find an ordinary singular term referring to an event in a

---

[4] Georg Henrik von Wright, *Norm and Action*, 23.

sentence like 'Caesar died' is properly explained by the fact that such sentences are existential and general with respect to events: we do not find a singular term referring to an event because there is none. But many philosophers, not doubting that 'Caesar died' refers to or describes an event, have confusedly concluded that the sentence *as a whole* refers to (or perhaps 'corresponds to') an event. As long ago as 1927, Frank Ramsey pointed out this error, and how to correct it; he described it as the error of conflating facts (which in his view are what sentences or propositions correspond to) and events.[5] And certainly there are difficulties, of a kind more general than we have indicated, with the idea that whole sentences refer to events. For suppose we agree, as I think we must, that the death of Scott is the same event as the death of the author of *Waverley*: then if sentences refer to events, the sentence 'Scott died' must refer to the same event as 'The author of *Waverley* died.' If we allow that substitution of singular terms in this way does not change the event referred to, then a short and persuasive argument will lead to the conclusion that all true sentences refer to the same event. And presumably only true sentences refer to an event; the conclusion may therefore be put: there is exactly one event. Since the argument is essentially the argument used by Frege to show that all sentences alike in truth-value must name the same thing, I spare you the details.[6]

The mistaken view that a sentence like 'Doris capsized the canoe yesterday' refers to a particular event, whether or not tied to the idea that it is the sentence as a whole that does the referring, is pretty sure to obliterate the difference between 'Doris capsized the canoe yesterday' and 'Doris' capsizing of the canoe occurred yesterday.' Yet without this distinction firm in our minds I do not believe we can make good sense of questions about the individuation of events and actions, for while the second sentence does indeed contain a singular description (the sentence as a whole meaning 'There is an event identical with the capsizing of the canoe yesterday by Doris'), the first sentence merely asserts the existence of at least one capsizing. If we are not alert to the difference, we are apt to ask wrongheaded questions like: if Jones apologized by saying 'I apologize', do 'Jones apologized' and 'Jones said "I apologize"' describes the same action? The right response is, I have urged, that neither sentence describes an action. We may then add,

[5] 'Facts and Propositions', 140, 141. Also see the reply to Martin in Essay 6.
[6] See Essays 6 and 7.

if we please, that at least one, or perhaps exactly one, action accounts for the truth of both sentences; but both sentences could be true although no apology by Jones was made by his saying, 'I apologize.'[7]

To see how not appreciating the generality in 'Jones apologized' can lead to mistakes about the individuation of events, consider a suggestion of Kim's.[8] Kim assumes that sentences such as 'Brutus killed Caesar' and 'Brutus stabbed Caesar' refer to events, and he asks under what conditions two such sentences describe or refer to the same event. He proposes the following criterion: two sentences are about the same event if they assert truly of the same particulars (i.e., substances) that the same properties (or relations) hold of them. Kim has a rather complicated doctrine of property identity, but it need not delay us since the point to be made depends only on a simple principle to which Kim agrees: properties differ if their extensions do. The effect is to substitute for what I think of as particular, dated events classes of such, and thus to make identities harder to come by. Where I would say the same event may make 'Jones apologized' and 'Jones said "I apologize"' true, Kim is committed to holding that these sentences describe different events. Nor can Kim allow that a stabbing is ever a killing, or the signing of a cheque the paying of a bill. He must also hold that if psychological predicates have no coextensive physical predicates, then no psychological event is identical with a physical event.

Kim recognizes these consequences of his criterion, and accepts them; but for reasons I find weak. He writes:

> Brutus' killing Caesar and Brutus' stabbing Caesar turn out, on the proposed criterion of event identity, to be different events, and similarly, 'Brutus killed Caesar' and 'Brutus stabbed Caesar' describe different events. Notice, however, that it is not at all absurd to say that Brutus' killing Caesar is *not the same as* Brutus' stabbing Caesar. Further, to explain Brutus' killing Caesar (why Brutus killed Caesar) is not the same as to explain Brutus' stabbing Caesar (why Brutus stabbed Caesar) . . . .[9]

---

[7] F. I. Dretske in 'Can Events Move?' correctly says that sentences do not refer to or describe events, and proposes that the expressions that do refer to events are the ones that can properly fill the blank in 'When did —— occur (happen, take place)?' This criterion includes (as it should) such phrases as 'the immersion of the paper' and 'the death of Socrates' but also includes (as it should not) 'a discoloration of the fluid'.

[8] In 'On the Psycho-Physical Identity Theory'. Essentially the same suggestion is made by Richard Martin in 'On Events and Event-Descriptions'.

[9] Op. cit., 232 (footnote).

Certainly Brutus had different reasons for stabbing Caesar than for killing him; we may suppose he went through a little piece of practical reasoning the upshot of which was that stabbing Caesar was a good way to do him in. But this reasoning was futile if, having stabbed Caesar, Brutus has a different action yet to perform (killing him). And explanation, like giving reasons, is geared to sentences or propositions rather than directly to what sentences are about: thus an explanation of why Scott died is not necessarily an explanation of why the author of *Waverley* died. Yet not even Kim wants to say the death of Scott is a different event from the death of the author of *Waverley*. I turn last to Kim's remark that it is not absurd to say that Brutus's killing Caesar is not the same as Brutus's stabbing Caesar. The plausibility in this is due, I think, to the undisputed fact that not all stabbings are killings. We are inclined to say: *this* stabbing might not have resulted in a death, so how can it be identical with the killing? Of course the death is not identical with the stabbing; it occurred later. But neither this nor the fact that some stabbings are not killings shows that this particular stabbing was not a killing. Brutus's stabbing of Caesar did result in Caesar's death; so it was in fact, though of course not necessarily, identical with Brutus's killing of Caesar.

Discussions of explanation may also suffer from confusion about how sentences are related to events. It is sometimes said, for example, that when we explain the occurrence of an event, we can do so only under one or another of its sentential descriptions. In so far as this remark reminds us of the essential intensionality of explanation, it is unexceptionable. But a mistake may lurk. If what we are to explain is why an avalanche fell on the village last week, we need to show that conditions were present adequate to produce *an* avalanche. It would be confused to say we have explained only an aspect of 'the real avalanche' if the reason for saying this lies in the fact that what was to be explained was itself general (for the explanandum contained no mention of a particular avalanche). We might instead have asked for an explanation of why *the* avalanche fell on the village last week. This is, of course, a harder task, for we are now asking not only why there was at least one avalanche, but also why there was not more than one. In a perfectly good sense the second explanation can be said to explain a particular event; the first cannot.

An associated point may be made about causal relations. Suppose it claimed that the lighting of this match was caused by the striking of the match. The inevitable comment (since the time of Mill anyway) is that the striking may have been *part* of the cause, but it was hardly sufficient for the lighting since it was also necessary for the match to be dry, that there be enough oxygen, etc. This comment is, in my opinion, confused. For since this match was dry, and was struck in enough oxygen, etc., the striking of this match was identical with the striking of a dry match in enough oxygen. How can one and the same event both be, and not be, sufficient for the lighting? In fact, it is not *events* that are necessary or sufficient as causes, but events as *described* in one way or another. It is true that we cannot infer, from the fact that the match was struck, and plausible causal laws, that the match lit; we can do better if we start with the fact that a dry match was struck in enough oxygen. It does not follow that more than the striking of this match was required to cause it to light.

Now that we have a clearer idea what it is like to have singular terms, say 'a' and 'b', that refer to events we may return to our original question when a sentence of the form '$a = b$' is true. Of course we cannot expect a general method for *telling* when such sentences are true. For suppose '$(\imath x)(Fx)$' describes some event. Letting 'S' abbreviate any sentence,

$$(\imath x)(Fx) = (\imath x)(Fx \ \&S)$$

is true just in case 'S' is true. Since 'S' is an arbitrary sentence, a general method for telling when identity sentences were true would have to include a method for telling when any sentence was true. What we want, rather, is a statement of necessary and sufficient conditions for identity of events, a satisfactory filling for the blank in:

If x and y are events, then $x = y$ if and only if____ .

Samples of answers (true or false) for other sorts of entities are: classes are identical if and only if they have exactly the same members; times are identical if and only if they are overlapped by exactly the same events; places are identical if and only if they are overlapped by exactly the same objects; material objects are identical if and only if they occupy exactly the same places at the same times. Can we do as well as this for events? Here follows a series of remarks that culminate in what I hope is a satisfactory positive answer.

(1) Many events are changes in a substance. If an event a is a change in some substance, then $a = b$ only if b is also a change in the same substance. Indeed, if $a = b$, every substance in which a is a change is identical with a substance in which b is a change. To touch on such necessary conditions of event-identity is to do little more than reflect on what follows if events really do exist; but that is to the present point. And of course we will not alter the event, if any, to which a description refers if in that description we substitute for the name or description of a substance another name or description of the same substance: witness the fact that the death of Scott is identical with the death of the author of *Waverley*. This is an example of a sufficient condition of identity.

We very often describe and identify events in terms of the objects to which they are in one way or another related. But it would be a mistake to suppose that, even for events that are naturally described as changes in an object, we *must* describe them (i.e., produce unique descriptions of them) by referring to the object. For in fact any predicate of any event may provide a unique description: if an event a is F, a may turn out also to be the only event that is F, in which case 'the event that is F' uniquely refers to a. One important way to identify events without explicit reference to a substance is by demonstrative reference: 'that shriek', 'that dripping sound', 'the next sonic boom'.

These last points are well made by Strawson.[10] Strawson also remarks that the possibilities for identifying events without reference to objects are limited, because, as he puts it, events do not provide 'a single, comprehensive and continuously usable framework' of reference of the kind provided by physical objects.[11] This claim is made by Strawson in support of a grander thesis, that events are conceptually dependent on objects. According to Strawson we could not have the idea of a birth or a death or a blow without the idea of an animal that is born or dies, or of an agent who strikes the blow.

I do not doubt that Strawson is right in this: most events are

---

[10] *Individuals*, 46ff. I am not sure, however, that Strawson distinguishes clearly among: pointing out an entity to someone; producing a unique description of an entity; producing a description guaranteed to be unique.

[11] Ibid., 53.

understood as changes in a more or less permanent object or substance. It even seems likely to me that the concept of an event depends in every case on the idea of a change in a substance, despite the fact that for some events it is not easy to say what substance it is that undergoes the change.

What does seem doubtful to me is Strawson's contention that while there is a conceptual dependence of the category of events on the category of objects, there is not a symmetrical dependence of the category of objects on the category of events. His principle argument may, I think, be not unfairly stated as follows: in a sentence like 'There is an event that is the birth of this animal' we refer to, or quantify over, events and objects alike. But we can, if we please, express exactly the same idea by saying, 'This animal was born' and here there is no reference to, or quantification over, events. We cannot in the same way eliminate the reference to the object.[12] This is supposed to show that objects are more fundamental than events.

A closely related argument of Strawson's is this: the sentence 'The blow which blinded John was struck by Peter' presupposes, for its truth, that John exists, that Peter exists, and that there is a striking of John by Peter. But the last presupposition may also be expressed simply by saying that Peter struck John, which does not treat the blow as an entity on a par with Peter and John. Strawson again concludes that events are dispensable in a sense in which objects are not.[13] It is hard to see how the evidence supports the conclusion.

If 'Peter struck John' and 'There was a striking of John by Peter' express the same presupposition, how can they require different ontologies? If 'This animal was born' and 'There is an event that is the birth of this animal' are genuine paraphrases one of the other, how can one of them be about a birth and the other not? The argument proves either too much or too little. If every context that seems to refer to, or to presuppose, events may be systematically rephrased so as not to refer to events, then this shows we do not need an ontology of events at all. On the other hand if some categories of sentence resist transformation into an eventless idiom, then the fact that we can apparently banish events from other areas cannot suffice to relegate events to a secondary status; indeed it

---

[12] Ibid., 51ff.   [13] Ibid., 200.

does not even serve to show that the sentences we know how to parse in superficially event-free terms are not about events. It was in fact in just this vein that I have been urging that we cannot give acceptable analyses of 'This animal was born' and 'Peter struck John' without supposing that there are such things as births and blows. In Strawson's view, if I understand him, 'The blow which blinded John was struck by Peter' entails 'Peter struck John.' But a theory about what these sentences mean that justifies the entailment must, or so I have argued, acknowledge an ontology of events. Thus if my interpretation of the evidence is correct, there is no reason to assign second rank to events; while if, contrary to what I have maintained, total reducibility is possible, then again events do not take a back seat, for there are no events.

In my view, a sentence like 'John struck the blow' is about two particulars, John and the blow. The distinction between singular terms and predicates is not abolished: rather, striking is predicated alike of John and of the blow. This symmetry in the treatment of substances and their changes reflects, I think, an underlying symmetry of conceptual dependence. Substances owe their special importance in the enterprise of identification to the fact that they survive through time. But the idea of survival is inseparable from the idea of surviving certain sorts of change—of position, size, shape, colour, and so forth. As we might expect, events often play an essential role in identifying a substance. Thus if we track down the author of *Waverley* or the father of Annette, it is by identifying an event, of writing, or of fathering. Neither the category of substance nor the category of change is conceivable apart from the other.[14]

(2) Should we say that events are identical only if they are in the same place? Of course if events have a location, same events have same locations; but here is a puzzle that may seem to cast a doubt on the project of assigning a clear location to events. Perhaps those events are easiest to locate that are obviously changes in some substance: we locate the event by locating the substance. But if one substance is part of another, a change in the first is a change in the second. Every substance is a part of the universe: hence every change is a change in the universe. It seems to follow that all simultaneous events have the same location. The error lies in the

---

[14] The same conclusion is reached by J. Moravscik, 'Strawson and Ontological Priority'.

assumption that if an event is a change in a substance, the location of the event is the entire space occupied by the substance. Rather, the location of the event at a moment is the location of the smallest part of the substance a change in which is identical with the event.

Does it make sense to assign a location to a mental event such as remembering that one has left a zipper open, deciding to schuss the headwall, or solving an equation? I think we do assign a location to such an event when we identify the person who remembered, decided, or solved: the event took place where the person was. Questions about the location of mental events are generally otiose because in identifying the event we have usually identified the person in whom the event was a change, so no interesting question about the location of the event remains that is not answered by knowing where the person was when the event occurred. When we do not know who the relevant person is, queries about the location of mental events are perfectly in order: 'Where was the infinitesimal calculus invented?'

Mental events (by which I mean events described in the mental vocabulary, whatever exactly that may be) are like many other sorts of events, and like material objects, in that we give their locations with no more accuracy than easy individuation (within the relevant vocabulary) demands. Aside from a few dubious cases, like pains, itches, pricks, and twitches, we have no reason to locate mental events more precisely than by identifying a person, for more than this would normally be irrelevant to individuation. Similarly, we uniquely identify a mountain by giving the latitude and longitude of its highest summit, and in one good sense this gives the location of the mountain. But a mountain is a material object, and so occupies more than a point; nevertheless convention decrees no formula for defining its boundaries.

An explosion is an event to which we find no difficulty in assigning a location, although again we may be baffled by a request to describe the total area. The following quotation from an article on locating earthquakes and underground explosions illustrates how smoothly we operate with the concept of the place of an event:

Information on the accuracy with which a seismic event can be located is not as complete as could be wished . . . . If data from stations distant from the event are used, it seems realistic to estimate that the site can be located within a circular area whose radius is about eight kilometers. Stations that are 500–2,000 kilometers from the event may give much larger errors....[15]

(3) No principle for the individuation of events is clearer or more certain than this, that if events are identical, they consume identical stretches of time. Yet even this principle seems to lead to a paradox.

Suppose I pour poison in the water tank of a spaceship while it stands on earth. My purpose is to kill the space traveller, and I succeed: when he reaches Mars he takes a drink and dies. Two events are easy to distinguish: my pouring of the poison, and the death of the traveller. One precedes the other, and causes it. But where does the event of my killing the traveller come in? The most usual answer is that my killing the traveller is identical with my pouring the poison. In that case, the killing is over when the pouring is. We are driven to the conclusion that I have killed the traveller long before he dies.

The conclusion to which we are driven is, I think, true, so coping with the paradox should take the form of reconciling us to the conclusion. First, we should observe that we may easily know that an event is a pouring of poison without knowing it is a killing, just as we may know that an event is the death of Scott with knowing it is the death of the author of *Waverley*. To describe an event as a killing is to describe it as an event (here an action) that caused a death, and we are not apt to describe an action as one that caused a death until the death occurs; yet it may be such an action before the death occurs. (And as it becomes more certain that a death will result from an action, we feel less paradox in saying, 'You have killed him'.)[16]

Directness of causal connection may also play a role. To describe the pouring as a killing is to describe it as the causing of a death; such a description loses cogency as the causal relation is attenuated. In general, the longer it takes for the effect to be registered, the more room there is for a slip, which is another way of saying, the less justification there is for calling the action alone the cause.

Finally, there may be a tendency to confuse events described (partly or wholly) in terms of terminal states and events described (partly or wholly) in terms of what they cause. Examples of the first sort are 'the rolling of the stone to the bottom of the hill' (which is

[15] E. C. Bullard, 'The Detection of Underground Explosions', 24.
[16] Harry Levin, *The Question of Hamlet*, 35, says in effect that the poisoned Hamlet, in killing the King, avenges, among other murders, his own. This he could not do if he had not already been murdered.

not over until the stone is at the bottom of the hill) or 'his painting the barn red' (not over until he has finished painting the barn red); examples of the second sort are 'the destruction of the crops by the flood' (over when the flood is, which may be finished before the crops are) and 'Jones' inviting Smith to the party' (which Jones does only if Smith gets invited, but has finished doing when he drops the card in the mail).[17]

It is a matter of the first importance that we may, and often do, describe actions and events in terms of their causal relations—their causes, their effects, or both. My poisoning of the victim must be an action that results in the victim being poisoned; my killing of the victim must be an action that results in the death of the victim; my murdering of the victim must be an action that results in the death of the victim and also an action that was caused, in part, by my desire for the victim's death. If I see that the cat is on the mat, my seeing must be caused, in part, by the cat's being on the mat. If I contract Favism, I must contract haemolytic anaemia as a consequence of eating, or otherwise coming in contact with, the Fava bean. And so forth. This tendency to identify events in terms of their causal relations has deep roots, as I shall suggest in a moment. But it should not lead to a serious difficulty about the dates of events.

(4) Do place and time together uniquely determine an event; that is, is it sufficient as well as necessary, if events are to be identical, that they occupy exactly the same time and the same place? This proposal was made (somewhat tentatively) by John Lemmon;[18] of course the same proposal has often been made for physical objects. I am uncertain both in the case of substances and in the case of events whether or not sameness of time and place is enough to insure identity. Doubt comes easily in the case of events, for it seems natural to say that two different changes can come over the whole of a substance at the same time. For example, if a metal ball becomes warmer during a certain minute, and during the same minute rotates through 35 degrees, must we say these are the same event? It would seem not; but there may be arguments the other way. Thus in the present instance it might be maintained that the warming of the ball

---

[17] I discuss this issue at greater length in Essay 3.

[18] E. J. Lemmon, 'Comments on D. Davidson's "The Logical Form of Action Sentences"'. Lemmon goes further, suggesting that '. . . we may invoke a version of the identity of indiscernables and identify events with *space-time zones*'. But even if there can be only one event that fully occupies a space-time zone, it would be wrong to say a space-time zone *is* a change or a cause (unless we want to alter the language).

during m is identical with the sum of the motions of the particles that constitute the ball during m; and so is the rotation. In the case of material objects it is perhaps possible to imagine two objects that in fact occupy just the same places at all times but are different because, though never separated, they are separable.

(5) We have not yet found a clearly acceptable criterion for the identity of events. Does one exist? I believe it does, and it is this: events are identical if and only if they have exactly the same causes and effects. Events have a unique position in the framework of causal relations between events in somewhat the way objects have a unique position in the spatial framework of objects. This criterion may seem to have an air of circularity about it, but if there is circularity it certainly is not formal. For the criterion is simply this: where x and y are events,

$(x = y$ if and only if $((z)$ $(z$ caused $x \leftrightarrow z$ caused $y)$ and (z) $(x$ caused $z \leftrightarrow y$ caused $z))$.

No identities appear on the right of the biconditional.

If this proposal is correct, then it is easy to appreciate why we so often identify or describe events in terms of their causes and effects. Not only are these the features that often interest us about events, but they are features guaranteed to individuate them in the sense not only of telling them apart but also of telling them together. It is one thing for a criterion to be correct, another for it to be useful. But there are certainly important classes of cases at least where the causal criterion appears to be the best we have. If we claim, for example, that someone's having a pain on a specific occasion is identical with a certain complex physiological event, the best evidence for the identity is apt to be whatever evidence we have that the pain had the same causes and the same effects as the physiological change. Sameness of cause and effect seems, in cases like this one, a far more useful criterion than sameness of place and time.[19]

Perhaps sameness of causal relations is the only condition always sufficient to establish sameness of events (sameness of location in space and time may be another). But this should not be taken to mean that the only way of establishing, or supporting, a claim that two events are identical is by giving causal evidence. On the contrary, logic alone, or logic plus physics, or almost anything else, may

[19] Thomas Nagel suggests the same criterion of the identity of events in 'Physicalism', 346.

help do the job, depending on the descriptions provided. What I do want to propose is that the causal nexus provides for events a 'comprehensive and continuously usable framework' for the identification and description of events analogous in many ways to the space-time coordinate system for material objects.

This paper may be viewed as an indirect defence of events as constituting a fundamental ontological category. A defence, because unless we can make sense of assertions and denials of identity we cannot claim to have made sense of the idea that events are particulars. Indirect, because it might be possible to make such needed sense, and to provide clear criteria for identity, and yet to have made no case at all for the need to posit events as an independent category. In other places I have tried to make good on the question of need; here I have not much more than summarized the arguments. But I have found that even those who are impressed with the arguments often have a residual doubt that centres on the apparent intractability of the question when events are identical.

I have tried to banish this doubt as far as I could. The results are not, it must be allowed, overwhelming. But how much should one expect? Can we do any better when it comes to giving criteria for individuating material objects? It should be noticed that the subject has been the individuation of events quite generally, not kinds of events. The analogous problem for material objects would be to ask for conditions of identity of equal generality. At this level, there is individuation without counting. We cannot answer the question, 'How many events occurred (since midnight, between Easter and Christmas)?' but neither can we answer the question, 'How many material objects are there (in the world, in this room)?' We do equally badly on counting classes, points, and intervals of time. Nor are there very good *formulas* for individuating in some of these cases, though we make good enough sense of assertions and denials of identity.

Individuation at its best requires sorts or kinds that give a principle for counting. But here again, events come out well enough: rings of the bell, major wars, eclipses of the moon, and performances of *Lulu* can be counted as easily as pencils, pots, and people. Problems can arise in either domain. The conclusion to be drawn, I think, is that the individuation of events poses no problems worse in principle than the problems posed by individuation of material objects; and there is as good reason to believe events exist.

9 *Events as Particulars*

Things change; but are there such things as changes? A pebble moves, an eland is born, a land slides, a star explodes. Are there, in addition to pebbles and stars, movements, births, landslides, and explosions? Our language encourages us in the thought that there are, by supplying not only appropriate singular terms, but the full apparatus of definite and indefinite articles, sortal predicates, counting, quantification, and identity-statements; all the machinery, it seems, of reference. If we take this grammar literally, if we accept these expressions and sentences as having the logical form they appear to have, then we are committed to an ontology of events as unrepeatable particulars ('concrete individuals'). It is to such events that we refer, or purport to refer, when we use descriptions like 'the death of Monteverdi', 'his second interview after the trial', 'the storm in the hills last night'; it is such events at least one of which we assert to exist when we assert 'There was a loud party in Gwen's apartment last week'; it is events of this kind one of which is said to be self-identical in 'His first attempt on the North Face was his last.' And it is events of this sort that we quantify over in 'All wars are preventable' and 'Not every lie has evil consequences.' That it is dated, particular events that seem to be required if such sentences are to be true is apparent from the principles of individuation implicit (for example) in the application of counting ('The third explosion was far more destructive than the first two', 'More than a third of all motorway accidents are caused by excess speed').

We have learned to be wary, however, of what the surface of language suggests, especially when it comes to ontology. In the present case, it is a striking fact that many sentences with what seems to intuition much the same subject matter as the sentences we have been quoting get along without obvious appeal to events. ('A pebble moved', 'the land slid', 'Monteverdi died', etc.). So events as particulars may not, after all, be basic to our understanding of the world. How can we tell?

We would be better placed to judge if we had a coherent, comprehensive account of the conditions under which our common beliefs (or believed-true sentences) are true. If we were in command of such a theory and that theory called for a domain of particular events, while our best efforts found no theory that worked as well without events, then we would have all the reason we could have for saying events exist; we would have all the reason we could have for saying that we do say that events exist.

We don't begin to have such a comprehensive theory, of course, but we can learn by trying. In Essays 6, 7, and 8 I proposed an analysis of sentences about events and actions that assumed a universe containing (*inter alia*) particular events. This analysis copes with a variety of problems in what seems to me an attractively simple way; and I know of no other theory that does as well. The theory won some supporters,[1] was criticized from various points of view,[2] and has acquired new rivals.[3] In the paper to which the present Essay is a response,[4] Roderick Chisholm emphasizes a problem I did not discuss, and to solve it he propounds an interesting and novel theory. He argues that his theory solves the new problem more efficiently than mine can, and deals with the original problems at least as well. In what follows I briefly compare the two approaches.

---

[1] John Lemmon and H.-N. Castañeda concurred in my criticisms of earlier theories, and accepted my proposal in general outline (though not in detail) in their comments on Essay 6. Roderick Chisholm wrote of that Essay, 'I think that my theory, on the whole, is consistent with [Davidson's] and indeed that these two theories may supplement each other' (113).

[2] For example, by Zeno Vendler when replying to Essay 7 in the *Journal of Philosophy*, 64 (1967), 704ff.; Charles Landesman, 'Actions as Universals'; and Richard Martin, in his reply to my comments on his paper, 'On Events and Event-Descriptions'. (My comments, somewhat edited, appear above under the heading 'Reply to Martin' following Essay 6; Martin's 'Reply' appears in *Fact and Existence*, 97–107.)

[3] Jaegwon Kim, 'On the Psycho-Physical Identity Theory'; Richard Montague, 'On the Nature of Certain Philosophical Entities' and 'Pragmatics'; Nicholas Rescher, 'Aspects of Action'; Richard Martin, 'On Events and Event-Descriptions'; the article by Landesman listed above.

[4] This Essay was part of a symposium; Chisholm's paper was 'Events and Propositions'. (For details, see the Introduction and Bibliography.)

What looms in the foreground in Chisholm's theory is 'the fact of *recurrence* . . . the fact that there are some things that recur, or happen more than once'. Perhaps this would be an example: last night I dropped a saucer of mud, and tonight I did it again (exactly the same thing happened). The 'it' of 'I did it again' looks for a reference, a thing that can recur.

One natural way to supply an appropriate entity would be to say that one and the same event, namely my dropping a saucer of mud, was instantiated both last night and tonight. Or we could say that one member of the class of my droppings of saucers of mud occurred last night, and another tonight. Chisholm observes that these analyses require two *sorts* of events, universals or classes on the one hand, and instances or members on the other; he hopes for greater ontological economy. His own solution is to invoke repeatable events that can be said to occur although there are, strictly speaking, no such things as their occurrences. The recurrence of my dropping a saucer of mud would be handled this way: my dropping a saucer of mud occurred both before and after the event of my not dropping a saucer of mud occurred. I find certain difficulties in this ingenious account, to which I shall (as one says) recur. But first I should like to make a more or less gratuitous remark about ontological parsimony, namely, what's so good about it? Clarity is desirable, but parsimony may or may not make for clarity. Of course, once one has a viable theory, it is interesting to learn that part only of the ontology one thought was needed will suffice; but such reduction must come after the provision of a working theory. One may also have intuitions that suggest what entities it is appropriate to call on in the interpretation of a given stretch of linguistic territory. I myself feel that to summon up classes or universals to explain 'I did it again' is using a cannon to shoot a mouse; so even if there are classes, I would prefer a theory that didn't call on them here. But again, economy is not the motive.

I sympathize, then, with Chisholm's goal of giving an account of talk about recurrence that does not require two categorially different kinds of event. But I think particular, unrepeatable events will do the job. One way is this: events have parts that are events, and the parts may be discontinuous temporally or spatially (think of a chess tournament, an argument, a war). Thus the *sum* of all my droppings of saucers of mud is a particular event, one of whose parts (which was a dropping of a saucer of mud by me) occurred last night; another such part occurred tonight. We need three events to carry this off, but they have the same ontological status.

Even if one allows only particular, unrepeatable events, then, it is possible to give a literal meaning to the claim that the same event occurs on two or more occasions. It is possible; but is this strange event-sum really what we refer to when we say, 'The same thing happened again'? A meeting can reconvene in another place, a play can continue after an intermission, a floating dice game can spring up again and again, and with new members. In these cases, we can talk of the same event *continuing*, perhaps after a pause. Is it plausible that when we say 'Jack and Jill got married in May, and Dolly and George did the same thing in June' we are saying that the event-sum of all marriages continued after a pause? Perhaps: 'The marrying resumed in June with Dolly and George.' But I confess that this seems strained, and the reason is, I think, that we normally do not require a single entity as reference to back every use of 'the same thing'. As Chisholm observes, our common talk is careless when it comes to identity: 'the same thing' often means 'something similar' or 'another'. 'Jones bought a leopard, and Smith bought the same thing' does not normally entail that there is a leopard both Jones and Smith bought. Analogously, 'Jones bought his wife a leopard, and Smith did the same thing' need not entail that there is a single action both performed. Smith and Jones did similar things: the character of the similarity is suggested, if not made explicit, by the context (did Smith buy his wife the same leopard that Jones bought his wife, or did Smith buy *Jones's* wife a leopard, etc.?) Recurrence may be no more than similar, but distinct, events following one after another.

If I am right, talk of the same event recurring no more requires an event that happens twice than talk of two tables having the same width requires there to be such a thing as the width both tables have. Of course this doesn't show Chisholm's view of events to be wrong; at best it shows that the theory of events as particulars is 'adequate to the fact of recurrence'. I turn now to what seem to me the difficulties in the way of developing Chisholm's theory.

Perhaps it will be agreed that in order to infer 'Meyer climbed the highest mountain in Africa' from 'Meyer climbed Kibo' we need no more than the premise 'Kibo is identical with the highest mountain in Africa.' But then no analysis of these sentences in terms of events can be correct that does not preserve the validity of the inference.

On my proposal, 'Meyer climbed Kibo' is analysed as saying that there exists an event that is a climbing of Kibo by Meyer: in symbols, '$(\exists x)$ (climbed (Kibo, Meyer, x))'. Clearly the desired inference is valid. According to Chisholm, however, we cannot infer the identity of Meyer's climbing of the highest mountain in Africa and Meyer's climbing of Kibo from the fact that Kibo is the highest mountain in Africa. Since 'Meyer climbed Kibo' is analysed, on Chisholm's account, as 'The climbing of Kibo by Meyer occurred', the inference with which we began is no longer shown to be valid. It would be possible, of course, simply to lay down additional rules of inference, but then these rules would have lost the connection with the conditions of truth of sentences that it is the business of systematic semantics to elicit. When we hazard a theory in response to the question, what kind of events, if any, must there be if our statements are to be true, we should be prepared to test the theory by considering what entailments between statements it sustains.

Chisholm does not give a convincing reason (it seems to me) for denying that the climbing of Kibo by Meyer is not identical with the climbing of the highest mountain in Africa by Meyer. Chisholm argues (to switch to his example) that Nixon's being in Washington is not the same event as Johnson's successor being in Washington, since we can say of the first event, but not of the second, that had Humphrey won, it would not have occurred. If this were a good argument, we could unhinge other true identity-statements: compare, 'We can say of Nixon, but not of Johnson's successor, that had Humphrey won, he would not have been president. Therefore Nixon is not Johnson's successor.'

Should Chisholm simply give up the attempt to distinguish between events under these conditions? The difficulty is that this course, given the fact that events for Chisholm are (speaking very roughly) entities that correspond to sentences, or sentence-like structures, is very apt to implement a proof that there are only two events altogether—the one that occurs, and the other. The trouble is a standard one, of admitting substitutivity of identity while denying extensionality. I am not saying it cannot be done, only that Chisholm has not spelled out a theory for doing it in the case of statements about events and actions.[5]

An adequate theory must give an account of adverbial modifica-

---

[5] I discuss this difficulty for theories that try to make events propositional in character in Essays 6 and 7.

tion; for example, the conditions under which (1) 'Sebastian strolled through the streets of Bologna at 2 a.m.' is said to be true must make clear why it entails (2) 'Sebastian strolled through the streets of Bologna.' If we analyse (1) as 'There exists an x such that Sebastian strolled x, x took place in the streets of Bologna, and x was going on at 2 a.m.' then the entailment is explained as logically parallel with (many cases of) adjectival modification; but this requires events as particulars. Chisholm analyses (1) as (1') 'There exists an x such that x is identical with the strolling of Sebastian, x occurred in the streets of Bologna, and x occurred at 2 a.m.' This does entail (2), but fails to entail 'Sebastian strolled'. (Chisholm's version of this last must be, 'There exists an x such that x is identical with the strolling of Sebastian and x occurred.') Chisholm might try modifying (1') to (1''): 'There exists an x such that x is identical with the strolling of Sebastian and x is in the streets of Bologna and x is at 2 a.m. and x occurred.' But (1'') shares with (1') the defect that it may be true when (1) is not: for example, if Sebastian strolled in Bologna at 8 p.m. and in Innsbruck at 2 a.m., but never in Bologna at 2 a.m. Indeed, (1'') is true if Sebastian ever took a stroll anywhere at any time. 'There exists an x such that x is the strolling of Sebastian in Bologna at 2 a.m. and x occurred' restores the connection between time and place, but at the cost of breaking the inferential link with (2).

It is not clear, then, how Chisholm's theory can cope with the problem of adverbial modification. If I am right, the following puzzle is at the bottom of much of the trouble. In order to account for recurrence, Chisholm says there is just one event of each kind: there is just one event that is the strolling of Sebastian, just one event that is the strolling of Sebastian in the streets of Bologna. In order to articulate the inference from (2) to 'Sebastian strolled', the event that is the strolling of Sebastian in the streets of Bologna must be complex: it is identical with the strolling of Sebastian, and it is in the streets of Bologna. But the strolling of Sebastian in the streets of Bologna cannot be identical with the strolling of Sebastian, since the second may recur and the first not. It is, however, a contradiction to assert that there is exactly one thing a that is both F and G, exactly one thing b that is F, and $a \neq b$.

The most breathtaking part of Chisholm's theory is his attempt to analyse what seem to be counting ('White failed the exams three times'), singularity ('The seventh round'), and quantification

('Every approach failed') without the expected entities to be counted, individuated, or quantified over. Central to the attempt are two concepts, that of the negation of an event, and that of an event occurring before another (or the same one) begins. Both concepts need further explanation. Since event negation differs from ordinary negation, we need an account of the semantic role of 'not-*p*' on the basis of the meaning of '*p*'. Chisholm understandably insists that the occurrence of not-*p* is not to be confused with the non-occurrence of *p*. But this does not answer the following question. If Sebastian strolled in the streets of Innsbruck, then surely he took a stroll that was not in the streets of Bologna. If we render this 'There exists an *x* such that *x* is the strolling of Sebastian, and *x* was not in the streets of Bologna and *x* occurred' then it contradicts the parallel analysis of 'Sebastian strolled in the streets of Bologna'; yet the unanalysed sentences are not contradictory.

The difficulty about '*p* occurs before *q* begins' is that it does not suggest how to analyse this locution in terms of 'before' and 'occurs'. But until this is done, the inference from '*p* occurs before *q* begins' to '*p* occurs' cannot be shown to be valid. Given individual events, the problem is easily solved: 'The moon rose after the sun set', for example, is rendered, 'There is an event *x* that is a rising of the moon, an event *y* that is a setting of the sun, and *x* came after *y*', and this plainly entails the occurrence of each event.

I think we must conclude that it is unclear whether a viable alternative to the theory of events as particulars can be worked out along the lines proposed by Chisholm. The difficulties I have raised depend in most cases on my having placed one or another interpretation on his suggestions; there may well be other ways of developing the theory that I have missed. Chisholm's approach treats event locutions much like mass terms ('snow', 'water', 'gold', 'wood'), and many of the troubles I find with his analysis of particular sentences are similar to the troubles that arise in the attempt to give a semantics for mass terms. There must be a right way to do it for mass terms (since 'Snow is white' is true if and only if snow is white), and if we knew this way, we might be able to adapt the method to apply to talk of events.[6] Chisholm's paper gives us an idea of how interesting it might be if such a theory could be successfully elaborated.

---

[6] I am encouraged in these reflections by unpublished work of John Wallace's that suggests an analysis of mass terms, and attempts to apply it to event sentences.

10 *Eternal* vs. *Ephemeral Events*

Roderick Chisholm's 'States of Affairs Again' removes some of the obstacles that separated us in an earlier exchange, and helps to sharpen the differences that remain.[1] Both of us think that there are events, but it is not clear that we agree about what events are; so joining the issue turns out to be tricky. Chisholm holds that timeless states of affairs (which may *occur* once, often, or never) are the ineluctable brand of event, but he has not denied that particular, unrepeatable events exist. I have not tried to show that states of affairs are supernumerary, only that particular events are not. What is directly in dispute, then, is not what sorts of entities exist. Where we disagree is on how entities are related to terms and sentences, on what kind of event must exist if certain typical sentences of ordinary language are to be true. If I am not mistaken, what emerges is a difference of opinion over how to do semantics, the study that relates language and ontology.

The dispute seems to me potentially instructive because Chisholm and I share a number of assumptions. He writes:

If (i) there is a sentence which seems to commit us to the existence of a certain object, (ii) we know the sentence to be true, and (iii) we can find no way of explicating or paraphrasing the sentence which will make clear to us that the truth of the sentence is compatible with the nonexistence of such an object, then it is more reasonable to suppose that there is such an object than it is not to suppose that there is such an object.[2]

I am in general agreement with this, and have said similar things in

---

[1] The earlier exchange was a symposium consisting of Chisholm's 'Events and Propositions' and Essay 9. 'States of Affairs Again' was a response to Essay 9.

[2] 'States of Affairs Again', 184.

connection with the present topic.[3] But I believe we can to advantage greatly sharpen the third condition, and when we do we will see that Chisholm's view of the relation between events and language is not satisfactory in its present form.

I direct attention first to some sentences that certainly seem to (and I think *do*) 'commit us to the existence of a certain object'.

(1) The explorer was in the cellar.
(2) The explosion was in the cellar.

The first of these sentences commits us to the existence of an explorer, a particular one picked out by the singular term 'the explorer'.[4] In a parallel way, the second sentence seems to commit us to the existence of a particular event. The impression should be deepened by the observation that (1) and (2) respectively entail:

(3) There was an explorer in the cellar.
(4) There was an explosion in the cellar.

What makes these inferences valid? The obvious answer for (1) and (3), an answer for which it is not easy to find plausible alternatives, is this: (1) is true if and only if there was exactly one explorer in the cellar, while (3) is true if and only if there was at least one. And if there was exactly one, there was at least one. This simple insight is made manifest if we rewrite (1) as 'The x such that x is an explorer was in the cellar' and (3) as 'There exists an x such that x is an explorer and x was in the cellar.' These paraphrases (avoidably awkward about tense) exhibit a syntactic element common to (1) and (3) (the predicate 'x is an explorer'), thus suggesting a formal rule of inference. What is essential, however, is not formality, but the justification of inference on the basis of conditions of truth. We need to see why what makes (1) true must make (3) true, and it is in accounting for this that we discover the need for an ontology of explorers. Or explosions, if we have our eye on (2) and (4).

With respect to relevant grammatical and logical matters, the

---

[3] For example, in the closing paragraphs of Essays 7 and 8. The former of these concludes, '. . . the assumption, ontological and metaphysical, that there are events, is one without which we cannot make sense of much of our most common talk . . . . I do not know any better, or further, way of showing what there is.'

[4] It would be irrelevant for present purposes to labour the fact that 'the explorer' will pick out a particular person only in a context of use. A sentence like (1) thus commits us to the existence of different objects on different occasions. Analogous remarks go for (2).

parallel between talk of explorers and talk of explosions is just about complete. As we have seen, explorers and explosions alike invite the definite and indefinite article; they also invite plural forms, universal quantification, counting, and identity statements. There is no more reason to think we can give a satisfactory account of the truth-conditions of sentences about explosions without invoking particular, dated events than there is to think we can give a satisfactory account of the truth-conditions of sentences about explorers without invoking particular, mortal bodies.

Perhaps it will be granted that nouns like 'explosion', 'death', 'collapse', 'disaster in the family', and so on, exhibit the same patterns of inference other common nouns do, and so we must expect the ontological assumptions that justify quantificational inference generally to apply here too. But how about the sentences that contain only the corresponding verbs? Why should we suppose there must be a particular event to verify

(5) The boiler exploded in the cellar.

when we see that there is no singular term to refer to an event, nor common noun to treat like a predicate true of events? There are good reasons for treating (5) as involving quantification over events, as having roughly the following logical form:

(6) There exists an x such that x was an explosion and x was in the cellar and x was of the boiler.

In this way we articulate the relations between the verb and the noun; we validate the inferences from (5) to (4) and from (5) to 'The boiler exploded.' In short, by treating (5) along the lines of (6) we explain why it is caught up in the network of entailments that intuition recognizes.

It is clear that the events over which the variables of quantification in sentences like (6) must range, the sort of explosions which we are taking (2) and (4) and (5) to be about, are particular, ephemeral events. If the boiler exploded twice, then there were at least two distinct explosions; in symbols:

$(\exists x)(\exists y)$ (x is an explosion & x is of the boiler & y is an explosion & y is of the boiler & $x \neq y$).

A very little reflection will show that *if* we treat 'x is an explosion' as a predicate true of events, and expressions like 'the explosion of the boiler' as referring terms, *then* we have chosen to include ephemeral events in our ontology.

Chisholm believes that such events will not serve to account for what he calls 'the fact of recurrence, . . . the fact that there are some things that recur, or happen more than once'.[5] Let us consider the situation. In Essay 9 I suggested that by treating sums of particular events as particular events we could give a literal meaning to the idea of the same event recurring. Although I am not persuaded that this device cannot be made to work, I shall not defend it here. For one thing, there is an air of unnatural contrivance about the idea. And for another, the important issue seems to me to lie not in the question of ontological economy, but in the question whether talk of 'recurrent' events requires us to suppose that there are timeless events that recur.

It might well be said that 'He takes a walk each Easter' is ambiguous, meaning either of the following:

(7) Each Easter he takes a walk.
(8) There is a walk he takes each Easter.

If the quantifications in (7) and (8) are interpreted literally, there must be two *kinds* of walk, one of the ephemeral sort he takes each Easter, the other the timeless sort that can be taken again and again. But while I think of (7) and (8) are certainly different in meaning, it is unclear that (8) calls for a repeatable event. For (8) may be paraphrased, 'There is a route along which, each Easter, he takes a walk.' Similarly, 'He danced the waltz eight times' seems properly rendered, 'The number of waltzes such that he danced them was eight.'

So far, there are no clear cases of sentences whose semantics are not satisfactorily, and even naturally, geared to the existence of particular events. But Chisholm now remarks[6] that I have not shown how to explicate sentences like:

(9) A certain event happened exactly twice.

Chisholm may intend (9) to give the form of a species of sentence he has in mind; the word 'certain' suggests this. Then a member of the species might be:

---

[5] See Chisholm, 'Events and Propositions', 15.
[6] See Chisholm, 'States of Affairs Again', 183.

(10) The exploding of the boiler happened exactly twice.

This is dubious English, but if it means anything I assume it means the same as 'The boiler exploded exactly twice', which is easily analysed in terms of particular explosions. But Chisholm may have in mind sentence (9) as it stands. I am inclined to think that if we were to hear (9) spoken, we would understand it only if the context provided a (perhaps tacit) 'namely' phrase after 'event'; and then the style of analysis used for (10) would apply. But if we are certain we understand (9) without ellipsis, then I would agree that it requires a timeless, undated entity for its analysis. This would put (9) on a par with a sentence like 'There is a certain fruit exactly two of which are on the table.' Those who wish to promote the existence of universals and other abstracta and eternalia have always insisted that we take such sentences as literal and central; those who are dubious about the entities are apt to think of these sentences as less than clear or less than literal. My heart is with the doubters, but I suspend judgement. A conditional claim remains: if we decide we must have eternal events to explain (9), we will still need the ephemeral ones to explain the simple inferences with which we began, just as we need particular fruit whether or not we decide that Kinds are also entities.

'Jones climbed the flagpole and Smith did the same thing' means the same as 'Jones climbed the flagpole and so did Smith' or, 'Jones and Smith climbed the flagpole'. There is no one action both performed; rather, the one predicate 'x is a climb up the flagpole' is true of both performances. As Russell says,

What is it that makes you recognize two strokes as two, and not as one thing repeated? ... When you hear the striking clock repeating closely similar noises ...[7]

My interest in the semantics of sentences about actions and events, their 'logical form', has from the start been spurred by the desire (among others) to make sense of statements like 'Action a is intentional under description d', for I found such language in writ-

---

[7] Bertrand Russell, *My Philosophical Development*, 163–4. Compare 'And suppose I have given my son a warning, "If you do anything like that again I'll spank you." Provided a future offence is serious enough, I will not be prepared to accept the plea, "But Papa, I did the very same thing again, not something like it."' Gareth Matthews gives credit for the example to Peter Geach. See Matthews, 'Mental Copies', 65.

ings I admired, and used it myself.[8] So I welcome Chisholm's challenge to explain, if I can, what I mean by this 'technical locution'.[9]

The explanation requires a clear separation of two problems. The first problem is to see how sentences that we loosely imagine are 'about' actions or events really refer to events, or contain descriptions of them. 'The setting of the Morning Star' and 'the setting of Venus' clearly enough have the form of descriptions, but how do sentences like 'The Morning Star set' refer to events? The answer that I have proposed is that this sentence has the form, 'There was a setting of the Morning Star' and so it involves quantification over events, but contains no description of one. On the other hand, 'The setting of the Morning Star took place' does contain a description and, unlike 'There was a setting of the Morning Star', is true only if the context picks out exactly one setting.[10] Once we accept the idea that there are events and actions, there is no difficulty in understanding the claim that two descriptions refer to the same one.

The second problem remains, for we have yet to elucidate sentences of the form, 'Action *a* is intentional under description *d*'. It should be said at once that this way of speaking is misleading. It suggests that '*Fa*' and '*a = b*' may be true and '*Fb*' false even though the position for a term after '*F*' is a referring position; but of course this would be a contradiction. Perhaps there is also the implication that references within statements of intention need to be relativized to a description. But the implication must be wrong, since in whatever sense such relativization is called for it is already present.

Part of the point then of speaking of an action or event 'under a description' is merely to make explicit the fact that some context is *intensional*. It is a context containing a position in which normal substitutivity of identity appears to break down (though whether it

---

[8] The motive comes to the surface in Essay 6 and Essay 9. A writing I admired: G. E. M. Anscombe, *Intention*; my own first use of the phrase is in Essay 1.

[9] Chisholm, 'States of Affairs Again', 187–8.

[10] This view is discussed at length in Essays 6, 7, and 8. Chisholm neglects the distinction between sentences like 'The setting of the Morning Star took place' and 'The Morning Star set' and in consequence misinterprets my view. Thus he says that according to me, 'If Jones climbed Mt. Monadnock, there is, or was, that event which was Jones climbing Mt. Monadnock.' ('States of Affairs Again', 180.) But on my view, 'Jones climbed Mt. Monadnock' is true if and only if there is at least one climbing of Mt. Monadnock by Jones; no single event is described or picked out. The same mistake mars Chisholm's discussion of my thesis on p. 182, lines 30ff. and the top of p. 187, in 'States of Affairs Again'.

actually breaks down depends on the final semantic analysis of such contexts), and in which the truth of a sentence can be affected by which of several (normally) coreferring descriptions is used.

The remainder of the point depends on the mixed nature of certain attributions of attitude or intent: examples are knowing, acting intentionally, perceiving that something is the case, remembering, and being pleased that something is the case. At one time Oedipus was pleased that he was married to Jocasta. It follows that he was married to Jocasta; and since Jocasta was his mother, that he was married to his mother. But he was not pleased that he was married to his mother. 'Oedipus was pleased that he was married to Jocasta' thus expresses a relation between Oedipus and Jocasta, but the truth of the sentence depends on Jocasta's being described in one way rather than another. It is natural, therefore, to think of the sentence as expressing a relation between Oedipus, Jocasta, and a certain description of Jocasta. In explaining why Oedipus was pleased, we might want to make explicit reference to the description of Jocasta under which he was pleased to be married to her.

The example concerns a relation between Oedipus, Jocasta, and a description of Jocasta, but a similar relation holds between Oedipus, Oedipus' marrying Jocasta, and a certain description of this event: the relation is expressed by 'Oedipus was pleased there was an event that was his marrying of Jocasta.' Intentional action provides other cases. It was intentional of Oedipus that there was an event that was his striking the old man at the crossroads. But though that event was identical with his striking his father, it was not intentional of Oedipus that there was an event identical with his striking his father. We may harmlessly compress the point by saying: the striking of the old man was intentional under one description but not under another. This does not mean the event did and did not have a certain property, but that the event, Oedipus, and a certain description, have a relation that does not obtain between the same event, Oedipus, and a different description.

None of this provides, or is meant to provide, an analysis of intensional idioms or attributions of attitudes. No detailed account of logical form has been suggested. The mention of 'descriptions' is obviously a gesture in the direction of ontology; but there can be no serious theory until we are told what descriptions are, and how attributions of attitude refer to them.

Chisholm is quite right, I think, in asking how it is possible to analyse intensional contexts 'in such a way that the result can be seen not to commit us to the existence of anything other than concrete particulars'.[11] 'Oedipus intentionally struck the old man' requires, for its truth, not only Oedipus and the old man (and, according to my view, a particular striking), but also some further entity, or entities: a proposition, state of affairs, meaning, description, sentence; pick your theory.[12]

Chisholm holds that states of affairs, as he understands them, will serve as objects of (some, at least, of) the attitudes, and also can be construed as events. I think both views are wrong, but I am not concerned here to dispute the first. The relevant point now is that events, in the sense in which we need them to explain ordinary inferences of the sort illustrated in the first pages of this paper, such events cannot serve as objects of the attitudes. Just as Oedipus' intention was directed not simply at the old man (since then it would equally and in the same way have been directed at his father), so it was not directed simply at the action of his striking the old man (since then it would equally and in the same way have been directed at his striking his father).

By looking at this matter from another angle, we can learn something about the individuation of events. Few would want to argue that the old man was not identical with Oedipus' father on the ground that Oedipus knew he had caused the death of the old man but did not know that he had caused the death of his father. But equally one should not reason from these facts that the death of the old man was not identical with the death of Oedipus' father.

Chisholm wants to persuade us that phrases like, 'Nixon's election' and 'the election of Johnson's successor' do not refer to the same event. However, all his examples of contexts in which such phrases cannot be substituted *salve veritate* are non-truth-functional, and so cannot resolve the point at issue.[13]

The conclusion of these last reflections is modest. I have not tried to show that Chisholm's states of affairs cannot be the objects of the attitudes, only that they cannot at the same time explain our ground level remarks and inferences about such events and actions as

---

[11] Chisholm, 'States of Affairs Again', 184.

[12] Elsewhere I have picked: see Essay 1 and 'On Saying That'. The extra entity, on my analysis, is an utterance, and hence, as it happens, a particular event.

[13] Chisholm, 'States of Affairs Again', 186.

explosions, killings and walks. In this respect, at least, they are on a par with physical objects.

I turn now to a final difficulty Chisholm finds in my theory. I think that 'Sebastian strolls in Bologna at 2 a.m.' may be analysed 'There is an event that Sebastian strolled, it was in Bologna, and it was at 2 a.m.' Chisholm concedes that this analysis provides a ground for certain inferences, but he adds,

... it leaves us with some perplexing questions. Consider that entity which, according to Davidson's analysis, Sebastian is said to stroll. Could some other person have strolled it? Could Sebastian have strolled it in Florence instead of Bologna? Or, had he not strolled it, could he have done something else with it instead? It would be unphilosophical, I think, to reject such questions—if one assumes that there really *is* a certain concrete thing that Sebastian strolls.[14]

It seems to me there are equally perplexing questions we can ask about Sebastian's nose. Could someone else have had it? Could it have been where his left ear is? If he had not been the possessor of his nose, could he have been in another relation to it? I do not think it is unphilosophical to ask such questions, but I do think it is a mistake to try to answer them out of context. Like many counterfactuals, the truth value, and indeed intelligibility, of these questions depends on background assumptions. Suppose that Sebastian's stroll was taken in response to orders from the commander of his secret society: each night someone, chosen by drawing a card, takes a stroll at 2 a.m. Then we might say that had the cards fallen out differently, another person would have taken that stroll. Or, if Sebastian takes a stroll each night at 2 a.m., we might say that if he had been in Florence that night instead of Bologna, then the stroll he did take in Bologna would have taken place in Florence.

The question whether Sebastian could have done something else with the stroll than stroll it is the oddest, because we do not normally use 'stroll' as a transitive verb. I originally treated it this way for the following reason:

There is an x such that x is a stroll and Sebastian took x' is more ornate than necessary, since there is nothing an agent can do with a stroll except take it; thus we may capture all there is with 'There is an x such that Sebastian strolled x'.[15]

This answers the question, whether an agent can do anything with a stroll except take it, in the negative. But of course this begs the

[14] Ibid., 182.   [15] Essay 8.

question, because unless the agent is the agent of that stroll, there are endless things he can do with a stroll besides take it: for one, he can make sure, by stacking the deck, that someone else takes it. For this reason, and others not directly relevant here, I now think the more ornate, and hence more articulate, version is to be preferred.

It remains to explain in slightly greater detail why Chisholm's states of affairs do not, as far as I can see, provide a satisfactory account of our ordinary talk about events and actions.

Chisholm's states of affairs are much like universals, although he denies that they are abstract, perhaps because they do not have instances. They are, however, timeless, and their existence is not a contingent matter, but is guaranteed by general principles (which I believe Chisholm would call logical). Since Chisholm does not mention that his principles entail that every event exists, I sketch the argument. Chisholm proposes the following two 'schemata':[16]

(A) $(\exists x)$ $((x$ consists in the fact that $p)$ & $(x$ occurs$)) \leftrightarrow p$
(B) $(\exists x)$ $((x$ consists in the fact that $p)$ & $\sim (x$ occurs$)) \leftrightarrow$ not-p

Since 'not-p' is the ordinary negation of any well-formed sentence, we may negate both sides of the second schema, and cancel the double negation. Then by appeal to the transitivity of the biconditional, we can combine the schemata to get:

(11) $(\exists x)$ $((x$ consists in the fact that $p)$ & $(x$ occurs$)) \leftrightarrow$
$\sim (\exists x)$ $((x$ consists in the fact that $p)$ & $\sim (x$ occurs$))$

Now suppose for a moment that the fact that p doesn't exist, i.e. that $\sim (\exists x)$ $(x$ consists in the fact that $p)$; then the right limb of (11) would be true, and so would the left. But the left side implies that the fact that p does exist. So the fact that p exists, whatever fact it may be.[17]

In Essay 9 I raised the question how well we understand

---

[16] Chisholm, 'States of Affairs Again', 181.

[17] That states of affairs can exist without occurring, is, of course, essential to their use in the analysis of attributions of attitude. Thus, even if Sebastian never strolls in Bologna, his strolling in Bologna may be the object of someone's belief. But there seems to be this difficulty. If I believe mistakenly that Sebastian strolled in Bologna, I do not merely believe such a stroll exists (in Chisholm's logically insured sense); I believe it occurred. The occurring is thus part of the object of my belief. Perhaps this trouble could be overcome by considering 'believes to occur' as a relation between believers and states of affairs, but then there would be the problem of relating this notion of belief to the usual one, whose objects are expressed by whole sentences.

Chisholm's distinction between the occurrence of not-*p* and the non-occurrence of *p*. In particular, I asked how Chisholm would cope with a sentence like 'Sebastian took a stroll that was not in Bologna.' Chisholm suggests that it be done as follows: 'There exists an *x* such that *x* consists in the fact that Sebastian strolled and *x* occurred in a place wholly other than Bologna.'[18] This solution leaves us with two problems. First, we do not know what the following well-formed sentence means: 'There is an *x* such that *x* is the strolling of Sebastian and *x* is not in the streets of Bologna' (since Chisholm agrees this doesn't mean that Sebastian took a stroll that was not in Bologna). And second, the solution works only on an ad hoc basis by finding some positive characteristic to replace a negative.

If I am right, the following derivation is authorized by Chisholm's principles. Suppose Jones climbs Mt. Monadnock every Easter. Then the fact that Jones climbs Mt. Monadnock on Easter recurs, which entails both that the fact occurs and that its negation (which Chisholm writes 'not-*p*') occurs.[19] So we have, as entailments of our original supposition:

(12) The fact that Jones climbs Mt. Monadnock on Easter occurs.
(13) The fact that Jones does not climb Mt. Monadnock on Easter occurs.

(in (13) I have been careful not to confuse the occurrence of a non-event with the non-occurrence of an event). But by schema (A), these are equivalent to:

(14) Jones climbs Mt. Monadnock on Easter.
(15) Jones does not climb Mt. Monadnock on Easter.

Are these, as they seem, contradictories? Well, by schema (B), (15) entails:

(16) The fact that Jones climbs Mt. Monadnock on Easter doesn't occur.

And (16) surely contradicts (12). My argument depends on assigning to the expression 'not-*p*' when it occurs in the discussion of recurrence the same meaning it has when it occurs in schema (B). If Chisholm decides that the expression means different things in the

---

[18] Chisholm, 'States of Affairs Again', 185.   [19] Essay 9.

two cases, then we must not only distinguish between the non-occurrence of p and the occurrence of not-p but also between two distinct kinds of non-events.

Schemata (A) and (B) raise further questions. Chisholm says that

(17) There occurs that event which is the strolling of Sebastian in Bologna at 2 a.m.

is an 'informal rendering' of

(18) ($\exists x$) (x consists in the fact that Sebastian strolls in Bologna at 2 a.m. and x occurs).

By schema (A), (18) is in turn equivalent to:

(19) Sebastian strolls in Bologna at 2 a.m.

Chisholm tells us that in the light of (A), (18) and (19) 'may be seen to entail each other'. I agree that (A) *says* they entail each other, and since (18) is not English, we may take Chisholm's word for it. But in what sense do we 'see' that they entail each other? Surely not in the sense in which we can sometimes be made to see that a rule of inference is valid. For this kind of insight comes from a recursive account of the conditions under which sentences are true, an account from which it follows that if sentences of one form are true, sentences of another form are. Chisholm has not given independent accounts of the truth-conditions of (18) and (19). But I assume that (18), which makes commitment to states of affairs explicit, is the form which Chisholm wants to knit into a serious theory, for if this cannot be done, it is unclear how the postulation of states of affairs can be justified.

I think Chisholm should say that (17) and (19) bear the same relation to (18): both are 'informal renderings' of an underlying form which is (in his view) made manifest by (18). If states of affairs really are required for the understanding of (18), they must equally be so for the understanding of (17) and (19). To put it differently, schemata (A) and (B), and indeed (18) itself, should be considered as part of a theory about the semantics (and ontology) of sentences like (19). The theory can be tested by asking whether (19), construed as the theory has it, entails, and is entailed by, the (intuitively) right sentences. It is no test to look to (17) and (18), since they are merely creatures of the theory.

We know that (19) is entailed by:

(20) The one stroll Sebastian took in Bologna was at 2 a.m.

How is this entailment shown to be valid by the analysis of (19) given by (18); how, in other words, does the idea that a state of affairs is referred to help us see that if (20) is true, (19) must be? Ad hoc rules of inference are no help here. The point of assuming an ontology and a deep analysis is to explain what seemed arbitrary. (20) says there was exactly one stroll of a certain sort; (19) says there was at least one such stroll: so of course (20) entails (19). But (18) and the sentence (A) would produce out of (20) *obscure* this fact, for each says there is exactly one state of affairs of a certain sort, but the sorts are different, and the theory does not relate them. By looking back at the discussion of (2) and (4), one can see how the appeal to particular events solves this problem.

Again, if (18) gives the correct analysis of (19), it ought to help us understand why (19) is entailed by, 'Every stroll Sebastian takes in Italy he takes in Bologna at 2 a.m.' and, 'Sebastian takes a stroll in Italy.' But (18) merely obstructs the inference; what is needed is clearly quantification over particular events.

A style of inference that is revealing involves substitutivity of identity. Since Bologna is the city with the oldest university in Europe, we ought to be able to infer from (19) that

(21) Sebastian strolls in the city with the oldest university in Europe at 2 a.m.

Normal practice makes the inference a matter of ordinary quantificational logic, but at the cost of treating 'x strolls in y at t' as a primitive three-place predicate, and hence blocking the inference (for example) to 'Sebastian strolls'. The theory I have proposed also validates the inference on ordinary quantificational grounds without blocking the inference to 'Sebastian strolls'; but to do this, it introduces an ontology of particular events. Chisholm's analysis again stands in the way of the inference. For him, (19) and (21) are to be understood as saying that a certain state of affairs occurs. In one case, that state of affairs is described as 'the fact that Sebastian strolls in Bologna at 2 a.m.'; in the other case, the description is the same except that 'the city with the oldest university in Europe' is substituted for 'Bologna'. If these descriptions were extensional with respect to the position occupied by the name or description of the place, then the inference might go through because then substitution of coreferring terms would yield descriptions of the same state of affairs.[20] But Chisholm denies that the state of affairs demanded by (19) is the same as that demanded by (21), as he must if states of affairs are to be the objects of attitudes. Thus the relation between the place referred to in the description of Sebastian's stroll, and what that description describes, remains a mystery, and the equivalence of (19) and (21) remains unexplained.

Chisholm has adjusted his account of what I call 'adverbial modification' to meet some of my earlier criticisms.[21] The problem is the now familiar one of giving a theory that will show why the inference, say, from (19) to 'Sebastian strolls at 2 a.m.' is valid. Chisholm's new solution is to state a rule of inference that is simply independent of his theory of states of affairs; like all rules of inference, it is purely syntactical in nature, and it is geared to sentences like (19) rather than (18). This rule cannot be used to support Chisholm's ontology, since it is unrelated to it. But neither can a rule of inference alone explain or show why an inference is valid in the desired sense, for it is the rule that needs to be shown valid. This can be done by proving that it leads from truths to truths in every case, which depends on showing exactly how the truth of the relevant sentences depends on their form, including, of course, the role of singular terms and variables. The rules of quantification theory can, we know, be shown valid in this sense, and the showing requires the identification of singular terms, and a specific decision about ontology. Chisholm has not offered us an alternative.

Chisholm apparently holds that such 'metaphysical' questions as whether there are events and states of affairs are irrelevant to logic.[22] I agree that these questions are metaphysical (since ontological), but it seems to me utterly wrong to divorce them from matters of entailment. The rules of logic aim at *describing* the valid inferences; they tell us that if some sentences are true, others must be. But they do not tell us *why*; justification must come from another source. Justification is supplied by a systematic account of truth which gives the conditions under which an arbitrary sentence is true; and doing this in a satisfactory way requires that various

---

[20] But to take this line is to risk identifying all states of affairs that occur. See Essay 6.

[21] The modified version is in Chisholm, 'States of Affairs Again', 181–2.

[22] Chisholm, 'States of Affairs Again', 181.

expressions be interpreted as referring to, or quantifying over, or being true of, certain entities.

If this is right, the logical relations between sentences provide the only real test of when our language commits us to the existence of entities. We have learned from Frege that it is only in the context of a sentence that a singular term comes to life; we should add that it is only in the context of a language that we can tell where the singular terms are. Chisholm thinks it is reasonable to assume that an entity exists if we can find no way, compatible with the non-existence of the entity, of paraphrasing or explicating a sentence we know to be true. I would add that the paraphrase or explication must spring from a coherent and systematic method of paraphrase that applies to the language as a whole, and that justifies the inferences we know to be valid. If we make this additional demand on an adequate theory, I think it is clear that whether or not Chisholm's states of affairs help us understand sentences about recurrence, and attributions of attitude, they do not yield a satisfactory account of the simplest sentences about events and actions.

PHILOSOPHY OF
PSYCHOLOGY

11 *Mental Events*

Mental events such as perceivings, rememberings, decisions, and actions resist capture in the nomological net of physical theory. How can this fact be reconciled with the causal role of mental events in the physical world? Reconciling freedom with causal determinism is a special case of the problem if we suppose that causal determinism entails capture in, and freedom requires escape from, the nomological net. But the broader issue can remain alive even for someone who believes a correct analysis of free action reveals no conflict with determinism. *Autonomy* (freedom, self-rule) may or may not clash with determinism; *anomaly* (failure to fall under a law) is, it would seem, another matter.

I start from the assumption that both the causal dependence, and the anomalousness, of mental events are undeniable facts. My aim is therefore to explain, in the face of apparent difficulties, how this can be. I am in sympathy with Kant when he says,

it is as impossible for the subtlest philosophy as for the commonest reasoning to argue freedom away. Philosophy must therefore assume that no true contradiction will be found between freedom and natural necessity in the same human actions, for it cannot give up the idea of nature any more than that of freedom. Hence even if we should never be able to conceive how freedom is possible, at least this apparent contradiction must be convincingly eradicated. For if the thought of freedom contradicts itself or nature . . . it would have to be surrendered in competition with natural necessity.[1]

Generalize human actions to mental events, substitute anomaly for freedom, and this is a description of my problem. And of course the connection is closer, since Kant believed freedom entails anomaly.

[1] *Fundamental Principles of the Metaphysics of Morals*, 75–6.

Now let my try to formulate a little more carefully the 'apparent contradiction' about mental events that I want to discuss and finally dissipate. It may be seen as stemming from three principles.

The first principle asserts that at least some mental events interact causally with physical events. (We could call this the Principle of Causal Interaction.) Thus for example if someone sank the *Bismarck*, then various mental events such as perceivings, notings, calculations, judgements, decisions, intentional actions, and changes of belief played a causal role in the sinking of the *Bismarck*. In particular, I would urge that the fact that someone sank the *Bismarck* entails that he moved his body in a way that was caused by mental events of certain sorts, and that this bodily movement in turn caused the *Bismarck* to sink.[2] Perception illustrates how causality may run from the physical to the mental: if a man perceives that a ship is approaching, then a ship approaching must have caused him to come to believe that a ship is approaching. (Nothing depends on accepting these as examples of causal interaction.)

Though perception and action provide the most obvious cases where mental and physical events interact causally, I think reasons could be given for the view that all mental events ultimately, perhaps through causal relations with other mental events, have causal intercourse with physical events. But if there are mental events that have no physical events as causes or effects, the argument will not touch them.

The second principle is that where there is causality, there must be a law: events related as cause and effect fall under strict deterministic laws. (We may term this the Principle of the Nomological Character of Causality.) This principle, like the first, will be treated here as an assumption, though I shall say something by way of interpretation.[3]

The third principle is that there are no strict deterministic laws on the basis of which mental events can be predicted and explained (the Anomalism of the Mental).

The paradox I wish to discuss arises for someone who is inclined to accept these three assumptions or principles, and who thinks they are inconsistent with one another. The inconsistency is not, of

---

[2] These claims are defended in Essays 1 and 3.

[3] In Essay 7, I elaborate on the view of causality assumed here. The stipulation that the laws be deterministic is stronger than required by the reasoning, and will be relaxed.

course, formal unless more premises are added. Nevertheless it is natural to reason that the first two principles, that of causal interaction and that of the nomological character of causality, together imply that at least some mental events can be predicted and explained on the basis of laws, while the principle of the anomalism of the mental denies this. Many philosophers have accepted, with or without argument, the view that the three principles do lead to a contradiction. It seems to me, however, that all three principles are true, so that what must be done is to explain away the appearance of contradiction; essentially the Kantian line.

The rest of this paper falls into three parts. The first part describes a version of the identity theory of the mental and the physical that shows how the three principles may be reconciled. The second part argues that there cannot be strict psychophysical laws; this is not quite the principle of the anomalism of the mental, but on reasonable assumptions entails it. The last part tries to show that from the fact that there can be no strict psychophysical laws, and our other two principles, we can infer the truth of a version of the identity theory, that is, a theory that identifies at least some mental events with physical events. It is clear that this 'proof' of the identity theory will be at best conditional, since two of its premises are unsupported, and the argument for the third may be found less than conclusive. But even someone unpersuaded of the truth of the premises may be interested to learn how they can be reconciled and that they serve to establish a version of the identity theory of the mental. Finally, if the argument is a good one, it should lay to rest the view, common to many friends and some foes of identity theories, that support for such theories can come only from the discovery of psychophysical laws.

I

The three principles will be shown consistent with one another by describing a view of the mental and the physical that contains no inner contradiction and that entails the three principles. According to this view, mental events are identical with physical events. Events are taken to be unrepeatable, dated individuals such as the particular eruption of a volcano, the (first) birth or death of a person, the playing of the 1968 World Series, or the historic utterance of the words, 'You may fire when ready, Gridley.' We can easily frame identity statements about individual events; examples (true or false) might be:

The death of Scott = the death of the author of *Waverley*;
The assassination of the Archduke Ferdinand = the event that started the First World War;
The eruption of Vesuvius in A.D. 79 = the cause of the destruction of Pompeii.

The theory under discussion is silent about processes, states, and attributes if these differ from individual events.

What does it mean to say that an event is mental or physical? One natural answer is that an event is physical if it is describable in a purely physical vocabulary, mental if describable in mental terms. But if this is taken to suggest that an event is physical, say, if some physical predicate is true of it, then there is the following difficulty. Assume that the predicate 'x took place at Noosa Heads' belongs to the physical vocabulary; then so also must the predicate 'x did not take place at Noosa Heads' belong to the physical vocabulary. But the predicate 'x did or did not take place at Noosa Heads' is true of every event, whether mental or physical.[4] We might rule out predicates that are tautologically true of every event, but this will not help since every event is truly describable either by 'x took place at Noosa Heads' or by 'x did not take place at Noosa Heads.' A different approach is needed.[5]

We may call those verbs mental that express propositional attitudes like believing, intending, desiring, hoping, knowing, perceiving, noticing, remembering, and so on. Such verbs are characterized by the fact that they sometimes feature in sentences with subjects that refer to persons, and are completed by embedded sentences in which the usual rules of substitution appear to break down. This criterion is not precise, since I do not want to include these verbs when they occur in contexts that are fully extensional ('He knows Paris,' 'He perceives the moon' may be cases), nor exclude them whenever they are not followed by embedded sentences. An alternative characterization of the desired class of men-

---

[4] The point depends on assuming that mental events may intelligibly be said to have a location; but it is an assumption that must be true if an identity theory is, and here I am not trying to prove the theory but to formulate it.

[5] I am indebted to Lee Bowie for emphasizing this difficulty.

tal verbs might be that they are psychological verbs as used when they create apparently nonextensional contexts.

Let us call a description of the form 'the event that is *M*' or an open sentence of the form 'event *x* is *M*' a *mental description* or a *mental open sentence* if and only if the expression that replaces '*M*' contains at least one mental verb essentially. (Essentially, so as to rule out cases where the description or open sentence is logically equivalent to one or not containing mental vocabulary.) Now we may say that an event is mental if and only if it has a mental description, or (the description operator not being primitive) if there is a mental open sentence true of that event alone. Physical events are those picked out by descriptions or open sentences that contain only the physical vocabulary essentially. It is less important to characterize a physical vocabulary because relative to the mental it is, so to speak, recessive in determining whether a description is mental or physical. (There will be some comments presently on the nature of a physical vocabulary, but these comments will fall far short of providing a criterion.)

On the proposed test of the mental, the distinguishing feature of the mental is not that it is private, subjective, or immaterial, but that it exhibits what Brentano called intentionality. Thus intentional actions are clearly included in the realm of the mental along with thoughts, hopes, and regrets (or the events tied to these). What may seem doubtful is whether the criterion will include events that have often been considered paradigmatic of the mental. Is it obvious, for example, that feeling a pain or seeing an after-image will count as mental? Sentences that report such events seem free from taint of nonextensionality, and the same should be true of reports of raw feels, sense data, and other uninterpreted sensations, if there are any.

However, the criterion actually covers not only the havings of pains and after-images, but much more besides. Take some event one would intuitively accept as physical, let's say the collision of two stars in distant space. There must be a purely physical predicate '*Px*' true of this collision, and of others, but true of only this one at the time it occurred. This particular time, though, may be pinpointed as the same time that Jones notices that a pencil starts to roll across his desk. The distant stellar collision is thus *the* event *x* such that *Px* and *x* is simultaneous with Jones's noticing that a pencil starts to roll across his desk. The collision has now been picked out by a mental description and must be counted as a mental event.

This strategy will probably work to show every event to be mental; we have obviously failed to capture the intuitive concept of the mental. It would be instructive to try to mend this trouble, but it is not necessary for present purposes. We can afford Spinozistic extravagance with the mental since accidental inclusions can only strengthen the hypothesis that all mental events are identical with physical events. What would matter would be failure to include bona fide mental events, but of this there seems to be no danger.

I want to describe, and presently to argue for, a version of the identity theory that denies that there can be strict laws connecting the mental and the physical. The very possibility of such a theory is easily obscured by the way in which identity theories are commonly defended and attacked. Charles Taylor, for example, agrees with protagonists of identity theories that the sole 'ground' for accepting such theories is the supposition that correlations or laws can be established linking events described as mental with events described as physical. He says, 'It is easy to see why this is so: unless a given mental event is invariably accompanied by a given, say, brain process, there is no ground for even mooting a general identity between the two.'[6] Taylor goes on (correctly, I think) to allow that there may be identity without correlating laws, but my present interest is in noticing the invitation to confusion in the statement just quoted. What can 'a given mental event' mean here? Not a particular, dated, event, for it would not make sense to speak of an individual event being 'invariably accompanied' by another. Taylor is evidently thinking of events of a given *kind*. But if the only identities are of kinds of events, the identity theory presupposes correlating laws.

One finds the same tendency to build laws into the statements of the identity theory in these typical remarks:

When I say that a sensation is a brain process or that lightning is an electrical discharge, I am using 'is' in the sense of strict identity . . . there are not two things: a flash of lightning and an electrical discharge. There is one thing, a flash of lightning, which is described scientifically as an electrical discharge to the earth from a cloud of ionized water molecules.[7]

[6] Charles Taylor, 'Mind–Body Identity, a Side Issue?', 202.

[7] J. J. C. Smart, 'Sensations and Brain Processes'. The quoted passages are on pages 163–5 of the reprinted version in *The Philosophy of Mind*, ed. V. C. Chappell (Englewood Cliffs, N.J., 1962). For another example, see David K. Lewis, 'An Argument for the Identity Theory'. Here the assumption is made explicit when Lewis takes events as universals (p. 17, footnotes 1 and 2). I do not suggest that Smart and

The last sentence of this quotation is perhaps to be understood as saying that for every lightning flash there exists an electrical discharge to the earth from a cloud of ionized water molecules with which it is identical. Here we have an honest ontology of individual events and can make literal sense of identity. We can also see how there could be identities without correlating laws. It is possible, however, to have an ontology of events with the conditions of individuation specified in such a way that any identity implies a correlating law. Kim, for example, suggests that *Fa* and *Gb* 'describe or refer to the same event' if and only if $a = b$ and the property of being F = the property of being G. The identity of the properties in turn entails that (x) $(Fx \leftrightarrow Gx)$.[8] No wonder Kim says:

If pain is identical with brain state B, there must be a concomitance between occurrences of pain and occurrences of brain state B. . . . Thus, a necessary condition of the pain-brain state B identity is that the two expressions 'being in pain' and 'being in brain state B' have the same extension. . . . There is no conceivable observation that would confirm or refute the identity but not the associated correlation.[9]

It may make the situation clearer to give a fourfold classification of theories of the relation between mental and physical events that emphasizes the independence of claims about laws and claims of identity. On the one hand there are those who assert, and those who deny, the existence of psychophysical laws; on the other hand there are those who say mental events are identical with physical and those who deny this. Theories are thus divided into four sorts: *nomological monism*, which affirms that there are correlating laws and that the events correlated are one (materialists belong in this category); *nomological dualism*, which comprises various forms of parallelism, interactionism, and epiphenomenalism; *anomalous dualism*, which combines ontological dualism with the general failure of laws correlating the mental and the physical (Cartesianism).

Lewis are confused, only that their way of stating the identity theory tends to obscure the distinction between particular events and kinds of events on which the formulations of my theory depends.

[8] Jaegwon Kim, 'On the Psycho-Physical Identity Theory', 231.

[9] Ibid., 227–8. Richard Brandt and Jaegwon Kim propose roughly the same criterion in 'The Logic of the Identity Theory'. They remark that on their conception of event identity, the identity theory 'makes a stronger claim than merely that there is a pervasive phenomenal-physical correlation', 518. I do not discuss the stronger claim.

And finally there is *anomalous monism*, which classifies the position I wish to occupy.[10]

Anomalous monism resembles materialism in its claim that all events are physical, but rejects the thesis, usually considered essential to materialism, that mental phenomena can be given purely physical explanations. Anomalous monism shows an ontological bias only in that it allows the possibility that not all events are mental, while insisting that all events are physical. Such a bland monism, unbuttressed by correlating laws or conceptual economies, does not seem to merit the term 'reductionism'; in any case it is not apt to inspire the nothing-but reflex ('Conceiving the *Art of the Fugue* was nothing but a complex neural event', and so forth).

Although the position I describe denies there are psychophysical laws, it is consistent with the view that mental characteristics are in some sense dependent, or supervenient, on physical characteristics. Such supervenience might be taken to mean that there cannot be two events alike in all physical respects but differing in some mental respect, or that an object cannot alter in some mental respect without altering in some physical respect. Dependence or supervenience of this kind does not entail reducibility through law or definition: if it did, we could reduce moral properties to descriptive, and this there is good reason to *believe* cannot be done; and we might be able to reduce truth in a formal system to syntactical properties, and this we *know* cannot in general be done.

This last example is in useful analogy with the sort of lawless monism under consideration. Think of the physical vocabulary as the entire vocabulary of some language L with resources adequate to express a certain amount of mathematics, and its own syntax. L' is L augmented with the truth predicate 'true-in-L', which is 'mental'. In L (and hence L') it is possible to pick out, with a definite description or open sentence, each sentence in the extension of the truth predicate, but if L is consistent there exists no predicate of syntax (of the 'physical' vocabulary), no matter how complex, that applies to all and only the true sentence of L. There can be no 'psychophysical law' in the form of a biconditional, '(x) (x is true-

[10] Anomalous monism is more or less explicitly recognized as a possible position by Herbert Feigl, 'The "Mental" and the "Physical"'; Sydney Shoemaker, 'Ziff's Other Minds'; David Randall Luce, 'Mind–Body Identity and Psycho-Physical Correlation'; Charles Taylor, op. cit., 207. Something like my position is tentatively accepted by Thomas Nagel, 'Physicalism', and endorsed by P. F. Strawson in *Freedom and the Will*, 63–7.

in-*L* if and only if *x* is φ)' where 'φ' is replaced by a 'physical' predicate (a predicate of *L*). Similarly, we can pick out each mental event using the physical vocabulary alone, but no purely physical predicate, no matter how complex, has, as a matter of law, the same extension as a mental predicate.

It should now be evident how anomalous monism reconciles the three original principles. Causality and identity are relations between individual events no matter how described. But laws are linguistic; and so events can instantiate laws, and hence be explained or predicted in the light of laws, only as those events are described in one or another way. The principle of causal interaction deals with events in extension and is therefore blind to the mental-physical dichotomy. The principle of the anomalism of the mental concerns events described as mental, for events are mental only as described. The principle of the nomological character of causality must be read carefully: it says that when events are related as cause and effect, they have descriptions that instantiate a law. It does not say that every true singular statement of causality instantiates a law.[11]

II

The analogy just bruited, between the place of the mental amid the physical, and the place of the semantical in a world of syntax, should not be strained. Tarski proved that a consistent language cannot (under some natural assumptions) contain an open sentence '*Fx*' true of all and only the true sentences of that language. If our analogy were pressed, then we would expect a proof that there can be no physical open sentence '*Px*' true of all and only the events having some mental property. In fact, however, nothing I can say about the irreducibility of the mental deserves to be called a proof; and the kind of irreducibility is different. For if anomalous monism is correct, not only can every mental event be uniquely singled out using only physical concepts, but since the number of events that falls under each mental predicate may, for all we know, be finite, there may well exist a physical open sentence coextensive with each

[11] The point that substitutivity of identity fails in the context of explanation is made in connection with the present subject by Norman Malcolm, Scientific Materialism and the Identity Theory', 123–4. Also see Essays 1 and 8.

mental predicate, though to construct it might involve the tedium of a lengthy and uninstructive alternation. Indeed, even if finitude is not assumed, there seems no compelling reason to deny that there could be coextensive predicates, one mental and one physical.

The thesis is rather that the mental is nomologically irreducible: there may be *true* general statements relating the mental and the physical, statements that have the logical form of a law; but they are not *lawlike* (in a strong sense to be described). If by absurdly remote chance we were to stumble on a nonstochastic true psychophysical generalization, we would have no reason to believe it more than roughly true.

Do we, by declaring that there are no (strict) psychophysical laws, poach on the empirical preserves of science—a form of *hubris* against which philosophers are often warned? Of course, to judge a statement lawlike or illegal is not to decide its truth outright; relative to the acceptance of a general statement on the basis of instances, ruling it lawlike must be a priori. But such relative apriorism does not in itself justify philosophy, for in general the grounds for deciding to trust a statement on the basis of its instances will in turn be governed by theoretical and empirical concerns not to be distinguished from those of science. If the case of supposed laws linking the mental and the physical is different, it can only be because to allow the possibility of such laws would amount to changing the subject. By changing the subject I mean here: deciding not to accept the criterion of the mental in terms of the vocabulary of the propositional attitudes. This short answer cannot prevent further ramifications of the problem, however, for there is no clear line between changing the subject and changing what one says on an old subject, which is to admit, in the present context at least, that there is no clear line between philosophy and science. Where there are no fixed boundaries only the timid never risk trespass.

It will sharpen our appreciation of the anomological character of mental–physical generalizations to consider a related matter, the failure of definitional behaviourism. Why are we willing (as I assume we are) to abandon the attempt to give explicit definitions of mental concepts in terms of behavioural ones? Not, surely, just because all actual tries are conspicuously inadequate. Rather it is because we are persuaded, as we are in the case of so many other forms of definitional reductionism (naturalism in ethics, instrumentalism and operationalism in the sciences, the causal theory of meaning, phenomenalism, and so on—the catalogue of philosophy's defeats), that there is system in the failures. Suppose we try to say, not using any mental concepts, what it is for a man to believe there is life on Mars. One line we could take is this: when a certain sound is produced in the man's presence ('Is there life on Mars?') he produces another ('Yes'). But of course this shows he believes there is life on Mars only if he understands English, his production of the sound was intentional, and was a response to the sounds as meaning something in English; and so on. For each discovered deficiency, we add a new proviso. Yet no matter how we patch and fit the nonmental conditions, we always find the need for an additional condition (provided he *notices, understands*, etc.) that is mental in character.[12]

A striking feature of attempts at definitional reduction is how little seems to hinge on the question of synonymy between definiens and definiendum. Of course, by imagining counterexamples we do discredit claims of synonymy. But the pattern of failure prompts a stronger conclusion: if we were to find an open sentence couched in behavioural terms and exactly coextensive with some mental predicate, nothing could reasonably persuade us that we had found it. We know too much about thought and behaviour to trust exact and universal statements linking them. Beliefs and desires issue in behaviour only as modified and mediated by further beliefs and desires, attitudes and attendings, without limit. Clearly this holism of the mental realm is a clue both to the autonomy and to the anomalous character of the mental.

These remarks apropos definitional behaviourism provide at best hints of why we should not expect nomological connections between the mental and the physical. The central case invites further consideration.

Lawlike statements are general statements that support counterfactual and subjunctive claims, and are supported by their instances. There is (in my view) no non-question-begging criterion of the lawlike, which is not to say there are no reasons in particular cases for a judgement. Lawlikeness is a matter of degree, which is not to deny that there may be cases beyond debate. And within limits set by the conditions of communication, there is room for much variation between individuals in the pattern of statements to which

---

[12] The theme is developed in Roderick Chisholm, *Perceiving*, Ch. 2.

various degrees of nomologicality are assigned. In all these respects nomologicality is much like analyticity, as one might expect since both are linked to meaning.

'All emeralds are green' is lawlike in that its instances confirm it, but 'all emeralds are grue' is not, for 'grue' means 'observed before time t and green, otherwise blue', and if our observations were all made before t and uniformly revealed green emeralds, this would not be a reason to expect other emeralds to be blue. Nelson Goodman has suggested that this shows that some predicates, 'grue' for example, are unsuited to laws (and thus a criterion of suitable predicates could lead to a criterion of the lawlike). But it seems to me the anomalous character of 'All emeralds are grue' shows only that the predicates 'is an emerald' and 'is grue' are not suited to one another: grueness is not an inductive property of emeralds. Grueness *is* however an inductive property of entities of other sorts, for instance of emerires. (Something is an emerire if it is examined before t and is an emerald, and otherwise is a sapphire.) Not only is 'All emerires are grue' entailed by the conjunction of a lawlike statements 'All emeralds are green' and 'All sapphires are blue,' but there is no reason, as far as I can see, to reject the deliverance of intuition, that it is itself lawlike.[13] Nomological statements bring together predicates that we know a priori are made for each other—know, that is, independently of knowing whether the evidence supports a connection between them. 'Blue', 'red', and 'green' are made for emeralds, sapphires, and roses; 'grue', 'bleen', and 'gred' are made for sapphalds, emerires, and emeroses.

The direction in which the discussion seems headed is this: mental and physical predicates are not made for one another. In point of lawlikeness, psychophysical statements are more like 'All emeralds are grue' than like 'All emeralds are green.'

Before this claim is plausible, it must be seriously modified. The fact that emeralds examined before t are grue not only is no reason to believe all emeralds are grue; it is not even a reason (if we know the time) to believe *any* unobserved emeralds are grue. But if an event of a certain mental sort has usually been accompanied by an

[13] The view is accepted by Richard C. Jeffrey, 'Goodman's Query', John R. Wallace, 'Goodman, Logic, Induction', and John M. Vickers, 'Characteristics of Projectible Predicates'. Goodman, in 'Comments', disputes the lawlikeness of statements like 'All emerires are grue.' I cannot see, however, that he meets the point of my 'Emeroses by Other Names'. This short paper is printed as an appendix to the present essay.

event of a certain physical sort, this often is a good reason to expect other cases to follow suit roughly in proportion. The generalizations that embody such practical wisdom are assumed to be only roughly true, or they are explicitly stated in probabilistic terms, or they are insulated from counterexample by generous escape clauses. Their importance lies mainly in the support they lend singular causal claims and related explanations of particular events. The support derives from the fact that such a generalization, however crude and vague, may provide good reason to believe that underlying the particular case there is a regularity that could be formulated sharply and without caveat.

In our daily traffic with events and actions that must be foreseen or understood, we perforce make use of the sketchy summary generalization, for we do not know a more accurate law, or if we do, we lack a description of the particular events in which we are interested that would show the relevance of the law. But there is an important distinction to be made within the category of the rude rule of thumb. On the one hand, there are generalizations whose positive instances give us reason to believe the generalization itself could be improved by adding further provisos and conditions stated in the same general vocabulary as the original generalization. Such a generalization points to the form and vocabulary of the finished law: we may say that it is a *homonomic* generalization. On the other hand there are generalizations which when instantiated may give us reason to believe there is a precise law at work, but one that can be stated only by shifting to a different vocabulary. We may call such generalizations *heteronomic*.

I suppose most of our practical lore (and science) is heteronomic. This is because a law can hope to be precise, explicit, and as exceptionless as possible only if it draws its concepts from a comprehensive closed theory. This ideal theory may or may not be deterministic, but it is if any true theory is. Within the physical sciences we do find homonomic generalizations, generalizations such that if the evidence supports them, we then have reason to believe they may be sharpened indefinitely by drawing upon further physical concepts: there is a theoretical asymptote of perfect coherence with all the evidence, perfect predictability (under the terms of the system), total explanation (again under the terms of the system). Or perhaps the ultimate theory is probabilistic, and the asymptote is less than perfection; but in that case there will be no better to be had.

Confidence that a statement is homonomic, correctible within its own conceptual domain, demands that it draw its concepts from a theory with strong constitutive elements. Here is the simplest possible illustration; if the lesson carries, it will be obvious that the simplification could be mended.

The measurement of length, weight, temperature, or time depends (among many other things, of course) on the existence in each case of a two-place relation that is transitive and asymmetric: warmer than, later than, heavier than, and so forth. Let us take the relation *longer than* as our example. The law or postulate of transitivity is this:

(L) $L(x, y)$ and $L(y,z) \rightarrow L(x,z)$

Unless this law (or some sophisticated variant) holds, we cannot easily make sense of the concept of length. There will be no way of assigning numbers to register even so much as ranking in length, let alone the more powerful demands of measurement on a ratio scale. And this remark goes not only for any three items directly involved in an intransitivity: it is easy to show (given a few more assumptions essential to measurement of length) that there is no consistent assignment of a ranking to any item unless (L) holds in full generality.

Clearly (L) alone cannot exhaust the import of 'longer than'—otherwise it would not differ from 'warmer than' or 'later than'. We must suppose there is some empirical content, however difficult to formulate in the available vocabulary, that distinguishes 'longer than' from the other two-place transitive predicates of measurement and on the basis of which we may assert that one thing is longer than another. Imagine this empirical content to be partly given by the predicate '$O(x,y)$'. So we have this 'meaning postulate':

(M) $O(x,y) \rightarrow L(x,y)$

that partly interprets (L). But now (L) and (M) together yield an empirical theory of great strength, for together they entail that there do not exist three objects a, b, and c such that $O(a,b)$, $O(b,c)$, and $O(c,a)$. Yet what is to prevent this happening if '$O(x,y)$' is a predicate we can ever, with confidence, apply? Suppose we *think* we observe an intransitive triad; what do we say? We could count (L) false, but then we would have no application for the concept of length. We could say (M) gives a wrong test for length; but then it is unclear what we thought was the *content* of the idea of one thing being longer than another. Or we could say that the objects under observation are not, as the theory requires, *rigid* objects. It is a mistake to think we are forced to accept some one of these answers. Concepts such as that of length are sustained in equilibrium by a number of conceptual pressures, and theories of fundamental measurement are distorted if we force the decision, among such principles as (L) and (M): analytic or synthetic. It is better to say the whole set of axioms, laws, or postulates for the measurement of length is partly constitutive of the idea of a system of macroscopic, rigid, physical objects. I suggest that the existence of lawlike statements in physical science depends upon the existence of constitutive (or synthetic a priori) laws like those of the measurement of length within the same conceptual domain.

Just as we cannot intelligibly assign a length to any object unless a comprehensive theory holds of objects of that sort, we cannot intelligibly attribute any propositional attitude to an agent except within the framework of a viable theory of his beliefs, desires, intentions, and decisions.

There is no assigning beliefs to a person one by one on the basis of his verbal behaviour, his choices, or other local signs no matter how plain and evident, for we make sense of particular beliefs only as they cohere with other beliefs, with preferences, with intentions, hopes, fears, expectations, and the rest. It is not merely, as with the measurement of length, that each case tests a theory and depends upon it, but that the content of a propositional attitude derives from its place in the pattern.

Crediting people with a large degree of consistency cannot be counted mere charity: it is unavoidable if we are to be in a position to accuse them meaningfully of error and some degree of irrationality. Global confusion, like universal mistake, is unthinkable, not because imagination boggles, but because too much confusion leaves nothing to be confused about and massive error erodes the background of true belief against which alone failure can be construed. To appreciate the limits to the kind and amount of blunder and bad thinking we can intelligibly pin on others is to see once more the inseparability of the question what concepts a person commands and the question what he does with those concepts in the way of belief, desire, and intention. To the extent that we fail to discover a coherent and plausible pattern in the attitudes and actions of others we simply forego the chance of treating them as persons.

The problem is not bypassed but given centre stage by appeal to explicit speech behaviour. For we could not begin to decode a man's sayings if we could not make out his attitudes towards his sentences, such as holding, wishing, or wanting them to be true. Beginning from these attitudes, we must work out a theory of what he means, thus simultaneously giving content to his attitudes and to his words. In our need to make him make sense, we will try for a theory that finds him consistent, a believer of truths, and a lover of the good (all by our own lights, it goes without saying). Life being what it is, there will be no simple theory that fully meets these demands. Many theories will effect a more or less acceptable compromise, and between these theories there may be no objective grounds for choice.

The heteronomic character of general statements linking the mental and the physical traces back to this central role of translation in the description of all propositional attitudes, and to the indeterminacy of translation.[14] There are no strict psychophysical laws because of the disparate commitments of the mental and physical schemes. It is a feature of physical reality that physical change can be explained by laws that connect it with other changes and conditions physically described. It is a feature of the mental that the attribution of mental phenomena must be responsible to the background of reasons, beliefs, and intentions of the individual. There cannot be tight connections between the realms if each is to retain allegiance to its proper source of evidence. The nomological irreducibility of the mental does not derive merely from the seamless nature of the world of thought, preference, and intention, for such interdependence is common to physical theory, and is compatible with there being a single right way of interpreting a man's attitudes without relativization to a scheme of translation. Nor is the irreducibility due simply to the possibility of many equally eligible schemes, for this is compatible with an arbitrary choice of one scheme relative to which assignments of mental traits are made. The

---

[14] The influence of W. V. Quine's doctrine of the indeterminacy of translation, as in Ch. 2 of *Word and Object*, is, I hope, obvious. In sect. 45 Quine develops the connection between translation and the propositional attitudes, and remarks that 'Brentano's thesis of the irreducibility of intentional idioms is of a piece with the thesis of indeterminacy of translation', 221.

point is rather that when we use the concepts of belief, desire, and
the rest, we must stand prepared, as the evidence accumulates, to
adjust our theory in the light of considerations of overall cogency:
the constitutive ideal of rationality partly controls each phase in the
evolution of what must be an evolving theory. An arbitrary choice
of translation scheme would preclude such opportunistic tempering
of theory; put differently, a right arbitrary choice of a translation
manual would be of a manual acceptable in the light of all possible
evidence, and this is a choice we cannot make. We must conclude, I
think, that nomological slack between the mental and the physical
is essential as long as we conceive of man as a rational animal.

III

The gist of the foregoing discussion, as well as its conclusion, will be
familiar. That there is a categorial difference between the mental
and the physical is a commonplace. It may seem odd that I say
nothing of the supposed privacy of the mental, or the special authority an agent has with respect to his own propositional attitudes, but
this appearance of novelty would fade if we were to investigate in
more detail the grounds for accepting a scheme of translation. The
step from the categorial difference between the mental and the
physical to the impossibility of strict laws relating them is less
common, but certainly not new. If there is a surprise, then, it will be
to find the lawlessness of the mental serving to help establish the
identity of the mental with that paradigm of the lawlike, the physical.

The reasoning is this. We are assuming, under the Principle of the
Causal Dependence of the Mental, that some mental events at least
are causes or effects of physical events; the argument applies only to
these. A second Principle (of the Nomological Character of Causality) says that each true singular causal statement is backed by a strict
law connecting events of kinds to which events mentioned as cause
and effect belong. Where there are rough, but homonomic, laws,
there are laws drawing on concepts from the same conceptual
domain and upon which there is no improving in point of precision
and comprehensiveness. We urged in the last section that such laws
occur in the physical sciences. Physical theory promises to provide a
comprehensive closed system guaranteed to yield a standardized,

unique description of every physical event couched in a vocabulary amenable to law.

It is not plausible that mental concepts alone can provide such a framework, simply because the mental does not, by our first principle, constitute a closed system. Too much happens to affect the mental that is not itself a systematic part of the mental. But if we combine this observation with the conclusion that no psychophysical statement is, or can be built into, a strict law, we have the Principle of the Anomalism of the Mental: there are no strict laws at all on the basis of which we can predict and explain mental phenomena.

The demonstration of identity follows easily. Suppose m, a mental event, caused p, a physical event; then, under some description m and p instantiate a strict law. This law can only be physical, according to the previous paragraph. But if m falls under a physical law, it has a physical description; which is to say it is a physical event. An analogous argument works when a physical event causes a mental event. So every mental event that is causally related to a physical event is a physical event. In order to establish anomalous monism in full generality it would be sufficient to show that every mental event is cause or effect of some physical event; I shall not attempt this.

If one event causes another, there is a strict law which those events instantiate when properly described. But it is possible (and typical) to know of the singular causal relation without knowing the law or the relevant descriptions. Knowledge requires reasons, but these are available in the form of rough heteronomic generalizations, which are lawlike in that instances make it reasonable to expect other instances to follow suit without being lawlike in the sense of being indefinitely refinable. Applying these facts to knowledge of identities, we see that it is possible to know that a mental event is identical with some physical event without knowing which one (in the sense of being able to give it a unique physical description that brings it under a relevant law). Even if someone knew the entire physical history of the world, and every mental event were identical with a physical, it would not follow that he could predict or explain a single mental event (so described, of course).

Two features of mental events in their relation to the physical—causal dependence and nomological independence—combine, then, to dissolve what has often seemed a paradox, the efficacy of thought and purpose in the material world, and their freedom from law. When we portray events as perceivings, rememberings, decisions and actions, we necessarily locate them amid physical happenings through the relation of cause and effect; but as long as we do not change the idiom that same mode of portrayal insulates mental events from the strict laws that can in principle be called upon to explain and predict physical phenomena.

Mental events as a class cannot be explained by physical science; particular mental events can when we know particular identities. But the explanations of mental events in which we are typically interested relate them to other mental events and conditions. We explain a man's free actions, for example, by appeal to his desires, habits, knowledge and perceptions. Such accounts of intentional behaviour operate in a conceptual framework removed from the direct reach of physical law by describing both cause and effect, reason and action, as aspects of a portrait of a human agent. The anomalism of the mental is thus a necessary condition for viewing action as autonomous. I conclude with a second passage from Kant:

It is an indispensable problem of speculative philosophy to show that its illusion respecting the contradiction rests on this, that we think of man in a different sense and relation when we call him free, and when we regard him as subject to the laws of nature. . . . It must therefore show that not only can both of these very well co-exist, but that both must be thought *as necessarily united* in the same subject. . . .[15]

APPENDIX: EMEROSES BY OTHER NAMES

Consider a hypothesis saying that everything that is examined before *t* and is an emerald (or else is a rose) is green if examined before *t* (or else is red); briefly:

H_1 All emeroses are gred

If H_1 is lawlike, it is a counterexample to Goodman's analysis in *Fact, Fiction and Forecast*, and one that would seem to cut pretty deep. Goodman's tests for deciding whether a statement is lawlike depend primarily on how well behaved its predicates are, taken one

---

[15] Op. cit., 76.

by one; thus for Goodman H_1 comes out doubly illegal. What H_1 suggests, however, is that it is a relation between the predicates that makes a statement lawlike, and it is not evident that this relation can be defined on the basis of the entrenchment of individual predicates.

But is H_1 lawlike? Recently Goodman has claimed it is not.[16] Here I consider whether he is right.

Let us pretend the following are true and lawlike:

H_2 All emeralds are green
H_3 All roses are red

Then H_1 is true, and we have good reason to believe it. Still, as Goodman points out, it does not follow that H_1 is lawlike, for it does not follow, from the fact that H_1 is entailed by hypotheses that are confirmed by their positive instances, that H_1 is confirmed by *its* positive instances.

Unless I am mistaken, the only reason Goodman gives for saying H_1 is not lawlike is contained in this remark: '... however true H_1 may be, it is unprojectible in that positive instances do not in general increase its credibility; emeralds found before t to be green do not confirm H_1' (328). Here the conclusion falls between comma and semicolon; what follows presumably gives the reason. The problem is to see how the reason supports the conclusion.

If positive instances were objects in the world, then the argument might be this: The positive instances of H_1 are gred emeroses, and if they are examined before t they are also green emeralds examined before t. But green emeralds examined before t do not tell us anything about the colour of roses examined after t. Unfortunately, if this were a good argument, it would also show that H_2 is not lawlike, for the positive instances of H_2 examined before t would be nothing but gred emeroses examined before t; and what can they tell us about the colour of emeralds after t?

In any case the assumption of the argument just examined is flatly at odds with clear indications in *Fact, Fiction and Forecast* (see p. 91, first edition, for example) that the positive instances of a hypothesis are sentences (or 'statements') immediately derivable from the

---

[16] Richard Jeffrey in 'Goodman's Query' and John Wallace in 'Goodman, Logic, Induction' generously mention me in connection with the difficulty apparently raised for Goodman by hypotheses like H_1, and Goodman responds in the first two pages of 'Comments'.

hypothesis by instantiation. The question whether H_1 is lawlike is then the question whether H_1 is confirmed by statements to the effect that this or that object is a gred emerose. Given this reading of 'positive instance', Goodman's remark quoted above seems to be a *non sequitur*: for how can the fact that H_1 is not confirmed by emeralds found before *t* to be green show that H_1 is not confirmed by statements that this or that object is a gred emerose?

The positive instances of H_1 do not mention time *t* any more than H_1 itself does. Nevertheless, an assumption of the discussion is that the objects described in the positive instances are actually observed before *t*, and perhaps a further assumption is that this fact is part of the background evidence against which the lawlike character of H_1 is to be judged. Given these assumptions, it is natural to suppose that the observer determines that an instance is positive by noting the time and observing that the object is a green emerald. But this supposition is adventitious, and may be false. I may know that at *t* a change will take place in the chemistry of my eye so that after *t* things that are red look green in normal light (before *t* green things look green); so, whether I know the time or not, I can tell by just looking that something is gred. Similarly I may be able to tell whether something is an emerose without knowing the time. Under these circumstances, it is hard to see why we would want to deny that H_1 is confirmed by its positive instances, i.e., that it is lawlike.

12 *Psychology as Philosophy*

Not all human motion is behaviour. Each of us in this room is moving eastward at about 700 miles an hour, carried by the diurnal rotation of the earth, but this is not a fact about our behaviour. When I cross my legs, the raised foot bobs gently with the beat of my heart, but I do not move my foot. Behaviour consists in things we do, whether by intention or not, but where there is behaviour, intention is relevant. In the case of actions, the relevance may be expressed this way: an event is an action if and only if it can be described in a way that makes it intentional. For example, a man may stamp on a hat, believing it is the hat of his rival when it is really his own. Then stamping on his own hat is an act of his, and part of his behaviour, though he did not do it intentionally. As observers we often describe the actions of others in ways that would not occur to them. This does not mean that the concept of intention has been left behind, however, for happenings cease to be actions or behaviour only when there is no way of describing them in terms of intention.[1]

These remarks merely graze a large subject, the relation between action and behaviour on the one hand, and intention on the other. I suggest that even though intentional action, at least from the point of view of description, is by no means all the behaviour there is, intention is conceptually central; the rest is understood and defined in terms of intention. If this is true, then any considerations which show that the intentional has traits that segregate it conceptually from other families of concepts (particularly physical concepts) will apply *mutatis mutandis* to behaviour generally. If the claim is mistaken, then the following considerations apply to psychology only to

---

[1] On the relation between intention and action, see Essay 3.

the extent that psychology employs the concepts of intention, belief, desire, hope, and other attitudes directed (as one says) upon propositions.

Can intentional human behaviour be explained and predicted in the same way other phenomena are? On the one hand, human acts are clearly part of the order of nature, causing and being caused by events outside ourselves. On the other hand, there are good arguments against the view that thought, desire and voluntary action can be brought under deterministic laws, as physical phenomena can. An adequate theory of behaviour must do justice to both these insights and show how, contrary to appearance, they can be reconciled. By evaluating the arguments against the possibility of deterministic laws of behaviour, we can test the claims of psychology to be a science like others (some others).

When the world impinges on a person, or he moves to modify his environment, the interactions can be recorded and codified in ways that have been refined by the social sciences and common sense. But what emerge are not the strict quantitative laws embedded in sophisticated theory that we confidently expect in physics, but irreducibly statistical correlations that resist, and resist in principle, improvement without limit. What lies behind our inability to discover deterministic psychophysical laws is this. When we attribute a belief, a desire, a goal, an intention or a meaning to an agent, we necessarily operate within a system of concepts in part determined by the structure of beliefs and desires of the agent himself. Short of changing the subject, we cannot escape this feature of the psychological; but this feature has no counterpart in the world of physics.[2]

The nomological irreducibility of the psychological means, if I am right, that the social sciences cannot be expected to develop in ways exactly parallel to the physical sciences, nor can we expect ever to be able to explain and predict human behaviour with the kind of precision that is possible in principle for physical phenomena. This does not mean there are any events that are in themselves undetermined or unpredictable; it is only events as described in the vocabulary of thought and action that resist incorporation into a closed deterministic system. These same events, described in appropriate physical terms, are as amenable to prediction and explanation as any.

---

[2] This claim, like several others in this Essay, is discussed at greater length in Essay 11.

I shall not argue here for this version of monism, but it may be worth indicating how the parts of the thesis support one another. Take as a first premise that psychological events such as perceivings, rememberings, the acquisition and loss of knowledge, and intentional actions are directly or indirectly caused by, and the causes of, physical events. The second premise is that when events are related as cause and effect, then there exists a closed and deterministic system of laws into which these events, when appropriately described, fit. (I ignore as irrelevant the possibility that microphysics may be irreducibly probabilistic.) The third premise, for which I shall be giving reasons, is that there are no precise psychophysical laws. The three premises, taken together, imply monism. For psychological events clearly cannot constitute a closed system; much happens that is not psychological, and affects the psychological. But is psychological events are causally related to physical events, there must, by premise two, be laws that cover them. By premise three, the laws are not psychophysical, so they must be purely physical laws. This means that the psychological events are describable, taken one by one, in physical terms, that is, they are physical events. Perhaps it will be agreed that this position deserves to be called *anomalous monism*: monism, because it holds that psychological events are physical events; anomalous, because it insists that events do not fall under strict laws when described in psychological terms.

My general strategy for trying to show that there are no strict psychophysical laws depends, first, on emphasizing the holistic character of the cognitive field. Any effort at increasing the accuracy and power of a theory of behaviour forces us to bring more and more of the whole system of the agent's beliefs and motives directly into account. But in inferring this system from the evidence, we necessarily impose conditions of coherence, rationality, and consistency. These conditions have no echo in physical theory, which is why we can look for no more than rough correlations between psychological and physical phenomena.

Consider our common-sense scheme for describing and explaining actions. The part of this scheme that I have in mind depends on the fact that we can explain why someone acted as he did by mentioning a desire, value, purpose, goal, or aim the person had, and a belief connecting the desire with the action to be explained. So, for example, we may explain why Achilles returned to the battle by saying he wished to avenge the death of Patroclus. (Given this much, we do not need to mention that he believed that by returning to the battle he could avenge the death of Patroclus.) This style of explanation has many variants. We may adumbrate explanation simply by expanding the description of the action: 'He is returning to battle with the intention of avenging the death of Patroclus.' Or we may more simply redescribe: 'Why is he putting on his armour?' 'He is getting ready to avenge Patroclus' death.' Even the answer, 'He just wanted to' falls into the pattern. If given in explanation of why Sam played the piano at midnight, it implies that he wanted to make true a certain proposition, that Sam play the piano at midnight, and he believed that by acting as he did, he would make it true.

A desire and a belief of the right sort may explain an action, but not necessarily. A man might have good reasons for killing his father, and he might do it, and yet the reasons not be his reasons in doing it (think of Oedipus). So when we offer the fact of the desire and belief in explanation, we imply not only that the agent had the desire and belief, but that they were *efficacious* in producing the action. Here we must say, I think, that causality is involved, i.e., that the desire and belief were causal conditions of the action. Even this is not sufficient, however. For suppose, contrary to the legend, that Oedipus, for some dark oedipal reason, was hurrying along the road intent on killing his father and, finding a surly old man blocking his way, killed him so he could (as he thought) get on with the main job. Then not only did Oedipus want to kill his father, and actually kill him, but his desire caused him to kill his father. Yet we could not say that in killing the old man he intentionally killed his father, nor that his reason in killing the old man was to kill his father.

Can we somehow give conditions that are not only necessary, but also sufficient, for an action to be intentional, using only such concepts as those of belief, desire and cause? I think not. The reason, very sketchily stated, is this. For a desire and a belief to explain an action in the right way, they must cause it in the right way, perhaps through a chain or process of reasoning that meets standards of rationality. I do not see how the right sort of causal process can be distinguished without, among other things, giving an account of how a decision is reached in the light of conflicting evidence and conflicting desires. I doubt whether it is possible to provide such an account at all, but certainly it cannot be done without using notions like evidence, or good reasons for believing, and these notions outrun those with which we began.[3]

What prevents us from giving necessary and sufficient conditions for acting on a reason also prevents us from giving serious laws connecting reasons and actions. To see this, suppose we had the sufficient conditions. Then we could say: whenever a man has such-and-such beliefs and desires, and such-and-such further conditions are satisfied, he will act in such-and-such a way. There are no serious laws of this kind. By a serious law, I mean more than a statistical generalization (the statistical laws of physics are serious because they give sharply fixed probabilities, which spring from the nature of the theory); it must be a law that, while it may have provisos limiting its application, allows us to determine in advance whether or not the conditions of application are satisfied. It is an error to compare a truism like 'If a man wants to eat an acorn omelette, then he generally will if the opportunity exists and no other desire overrides' with a law that says how fast a body will fall in a vacuum. It is an error, because in the latter case, but not the former, we can tell in advance whether the condition holds, and we know what allowance to make if it doesn't. What is needed in the case of action, if we are to predict on the basis of desires and beliefs, is a quantitative calculus that brings all relevant beliefs and desires into the picture. There is no hope of refining the simple pattern of explanation on the basis of reasons into such a calculus.

Two ideas are built into the concept of acting on a reason (and hence, the concept of behaviour generally): the idea of cause and the idea of rationality. A reason is a rational cause. One way rationality is built in is transparent: the cause must be a belief and a desire in the light of which the action is reasonable. But rationality also enters more subtly, since the way desire and belief work to cause the action must meet further, and unspecified, conditions. The advantage of this mode of explanation is clear: we can explain behaviour without having to know too much about how it was caused. And the cost is appropriate: we cannot turn this mode of explanation into something more like science.

Explanation by reasons avoids coping with the complexity of causal factors by singling out one, something it is able to do by omitting to provide, within the theory, a clear test of when the

[3] See Essay 4.

antecedent conditions hold. The simplest way of trying to improve matters is to substitute for desires and beliefs more directly observable events that may be assumed to cause them, such as flashing lights, punishments and rewards, deprivations, or spoken commands and instructions. But perhaps it is now obvious to almost everyone that a theory of action inspired by this idea has no chance of explaining complex behaviour unless it succeeds in inferring or constructing the pattern of thoughts and emotions of the agent.

The best, though by no means the only, evidence for desires and beliefs is action, and this suggests the possibility of a theory that deals directly with the relations between actions, and treats wants and thoughts as theoretical constructs. A sophisticated theory along these lines was proposed by Frank Ramsey.[4] Ramsey was primarily interested in providing a foundation in behaviour for the idea that a person accords one or another degree of credence to a proposition. Ramsey was able to show that if the pattern of an individual's preferences or choices among an unlimited set of alternatives meets certain conditions, then that individual can be taken to be acting so as to maximize expected utility, that is, he acts as if he assigns values to the outcomes on an interval scale, judges the plausibility of the truth of propositions on a ratio scale, and chooses the alternative with the highest computed expected yield.

Ramsey's theory suggests an experimental procedure for disengaging the roles of subjective probability (or degree of belief) and subjective value in choice behaviour. Clearly, if it may be assumed that an agent judges probabilities in accord with frequencies or so-called objective probabilities, it is easy to compute from his choices among gambles what his values are; and similarly one can compute his degree of belief in various propositions if one can assume that his values are, say, linear in money. But neither assumption seems justified in advance of evidence, and since choices are the resultant of both factors, how can either factor be derived from choices until the other is known? Here, in effect, is Ramsey's solution: we can tell that a man judges an event as likely to happen as not if he doesn't care whether an attractive or an

---

[4] 'Truth and Probability'. Ramsey's theory, in a less interesting form, was later, and independently, rediscovered by von Neumann and Morgenstern, and is sometimes called a theory of decision under uncertainty, or simply decision theory, by economists and psychologists.

unattractive outcome is tied to it, if he is indifferent, say, between these two options:

	Option 1	Option 2
If it rains you get:	$1,000	a kick
If it doesn't rain:	a kick	$1,000

Using this event with a subjective probability of one half, it is possible to scale values generally, and using these values, to scale probabilities.

In many ways, this theory takes a long step towards scientific respectability. It gives up trying to explain actions one at a time by appeal to something more basic, and instead postulates a pattern in behaviour from which beliefs and attitudes can be inferred. This simultaneously removes the need for establishing the existence of beliefs and attitudes apart from behaviour, and takes into systematic account (as a construct) the whole relevant network of cognitive and motivational factors. The theory assigns numbers to measure degrees of belief and desire, as is essential if it is to be adequate to prediction, and yet it does this on the basis of purely qualitative evidence (preferences or choices between pairs of alternatives). Can we accept such a theory of decision as a scientific theory of behaviour on a par with a physical theory?

Well, first we must notice that a theory like Ramsey's has no predictive power at all unless it is assumed that beliefs and values do not change over time. The theory merely puts restrictions on a temporal cross-section of an agent's dispositions to choose. If we try experimentally to test the theory, we run into the difficulty that the testing procedure disturbs the pattern we wish to examine. After spending several years testing variants of Ramsey's theory on human subjects, I tried the following experiment (with Merrill Carlsmith). Subjects made all possible pairwise choices within a small field of alternatives, and in a series of subsequent sessions, were offered the same set of options over and over. The alternatives were complex enough to mask the fact of repetition, so that subjects could not remember their previous choices, and pay-offs were deferred to the end of the experiment so that there was no normal learning or conditioning. The choices for each session and each subject were then examined for inconsistencies—cases where someone had chosed *a* over *b*, *b* over *c*, and *c* over *a*. It was found that as time went on, people became steadily more consistent;

intransitivities were gradually eliminated; after six sessions, all subjects were close to being perfectly consistent. This was enough to show that a static theory like Ramsey's could not, even under the most carefully controlled conditions, yield accurate predictions: merely making choices (with no reward or feedback) alters future choices. There was also an entirely unexpected result. If the choices of an individual over all trials were combined, on the assumption that his 'real' preference was for the alternative of a pair he chose most often, then there were almost no inconsistencies at all. Apparently, from the start there were underlying and consistent values which were better and better realized in choice. I found it impossible to construct a formal theory that could explain this, and gave up my career as an experimental psychologist.

Before drawing a moral from this experiment, let me return to Ramsey's ingenious method for abstracting subjective values and probabilities simultaneously from choice behaviour. Application of the theory depends, it will be remembered, on finding a proposition with a certain property: it must be such that the subject does not care whether its truth or its falsity is tied to the more attractive of two outcomes. In the context of theory, it is clear that this means, *any* two outcomes. So, if the theory is to operate at all, if it is to be used to measure degrees of belief and the relative force of desire, it is first necessary that there be a proposition of the required sort. Apparently, this is an empirical question; yet the claim that the theory is true is then a very sweeping empirical claim. If it is ever correct, according to the theory, to say that for a given person a certain event has some specific subjective probability, it must be the case that a detailed and powerful theory is true concerning the pattern of that person's choice behaviour. And if it is ever reasonable to assert, for example, that one event has a higher subjective probability than another for a given person, then there must be good reason to believe that a very strong theory is true rather than false.

From a formal point of view, the situation is analogous to fundamental measurement in physics, say of length, temperature, or mass. The assignment of numbers to measure any of these assumes that a very tight set of conditions holds. And I think that we can treat the cases as parallel in the following respect. Just as the satisfaction of the conditions for measuring length or mass may be viewed as constitutive of the range of application of the sciences that employ these measures, so the satisfaction of conditions of consistency and rational coherence may be viewed as constitutive of the range of applications of such concepts as those of belief, desire, intention and action. It is not easy to describe in convincing detail an experiment that would persuade us that the transitivity of the relation of *heavier than* had failed. Though the case is not as extreme, I do not think we can clearly say what should convince us that a man at a given time (without change of mind) preferred *a* to *b*, *b* to *c*, and *c* to *a*. The reason for our difficulty is that we cannot make good sense of an attribution of preference except against a background of coherent attitudes.

The significance of the experiment I described a page or so back is that it demonstrates how easy it is to interpret choice behaviour so as to give it a consistent and rational pattern. When we learn that apparent inconsistency fades with repetition but no learning, we are apt to count the inconsistency as merely apparent. When we learn that frequency of choice may be taken as evidence for an underlying consistent disposition, we may decide to write off what seem to be inconsistent choices as failures of perception or execution. My point is not merely that the data are open to more than one interpretation, though this is obviously true. My point is that if we are intelligibly to attribute attitudes and beliefs, or usefully to describe motions as behaviour, then we are committed to finding, in the pattern of behaviour, belief, and desire, a large degree of rationality and consistency.

A final consideration may help to reinforce this claim. In the experiments I have been describing, it is common to offer the subject choices verbally, and for him to respond by saying what he chooses. We assume that the subject is choosing between the alternatives described by the experimenter, i.e. that the words used by subject and experimenter have the same interpretation. A more satisfying theory would drop the assumption by incorporating in decision theory a theory of communication. This is not a peripheral issue, because except in the case of the most primitive beliefs and desires, establishing the correctness of an attribution of belief or desire involves much the same problems as showing that we have understood the words of another. Suppose I offer a person an apple and a pear. He points to the apple, and I record that he has chosen the apple. By describing his action in this way, I imply that he intended to point to the apple, and that by pointing he intended to

238 *Philosophy of Psychology*

indicate his choice. I also imply that he believed he was choosing an apple. In attributing beliefs we can make very fine distinctions, as fine as our language provides. Not only is there a difference between his believing he is choosing an apple and his believing he is choosing a pear. There is even a difference between his believing he is choosing the best apple in the box and his believing he is choosing the largest apple, and this can happen when the largest is the best.

All the distinctions available in our language are used in the attribution of belief (and desire and intention); this is perhaps obvious from the fact that we can attribute a belief by putting any declarative sentence after the words, 'He believes that'. There is every reason to hold, then, that establishing the correctness of an attribution of belief is no easier than interpreting a man's speech. But I think we can go further, and say that the problems are identical. Beliefs cannot be ascertained in general without command of a man's language; and we cannot master a man's language without knowing much of what he believes. Unless someone could talk with him, it would not be possible to know that a man believed Fermat's last theorem to be true, or that he believed Napoleon had all the qualities of a great general.

The reason we cannot understand what a man means by what he says without knowing a good deal about his beliefs is this. In order to interpret verbal behaviour, we must be able to tell when a speaker holds a sentence he speaks to be true. But sentences are held to be true partly because of what is believed, and partly because of what the speaker means by his words. The problem of interpretation therefore is the problem of abstracting simultaneously the roles of belief and meaning from the pattern of sentences to which a speaker subscribes over time. The situation is like that in decision theory: just as we cannot infer beliefs from choices without also inferring desires, so we cannot decide what a man means by what he says without at the same time constructing a theory about what he believes.

In the case of language, the basic strategy must be to assume that by and large a speaker we do not yet understand is consistent and correct in his beliefs—according to our own standards, of course. Following this strategy makes it possible to pair up sentences the speaker utters with sentences of our own that we hold true under like circumstances. When this is done systematically, the result is a method of translation. Once the project is under way, it is possible, and indeed necessary, to allow some slack for error or difference of opinion. But we cannot make sense of error until we have established a base of agreement.

The interpretation of verbal behaviour thus shows the salient features of the explanation of behaviour generally: we cannot profitably take the parts one by one (the words and sentences), for it is only in the context of the system (language) that their role can be specified. When we turn to the task of interpreting the pattern, we notice the need to find it in accord, within limits, with standards of rationality. In the case of language, this is apparent, because understanding it is *translating* it into our own system of concepts. But in fact the case is no different with beliefs, desires, and actions.

The constitutive force in the realm of behaviour derives from the need to view others, nearly enough, as like ourselves. As long as it is behaviour and not something else we want to explain and describe, we must warp the evidence to fit this frame. Physical concepts have different constitutive elements. Standing ready, as we must, to adjust psychological terms to one set of standards and physical terms to another, we know that we cannot insist on a sharp and law-like connection between them. Since psychological phenomena do not constitute a closed system, this amounts to saying they are not, even in theory, amenable to precise prediction or subsumption under deterministic laws. The limit thus placed on the social sciences is set not by nature, but by us when we decide to view men as rational agents with goals and purposes, and as subject to moral evaluation.

COMMENTS AND REPLIES

Essay 12 was delivered (in slightly altered form) at a conference on the Philosophy of Psychology organized by the Royal Institute of Philosophy and held at the University of Kent at Canterbury in September, 1971. The proceedings were published in *Philosophy of Psychology*, edited by Stuart Brown, and contained comments on my remarks by the chairman of the session, Professor Richard Peters, and questions asked by Mr Robin Attfield, Professor Les Holborrow, and Professor Robert Solomon. My replies follow.

Reply to Peters. The conclusion defended in this paper is a familiar one, and is shared by many philosophers and, probably, psychologists. The position might be put this way: the study of human action, motives, desires, beliefs, memory, and learning, at least so far as these are logically tied to the so-called 'propositional attitudes', cannot employ the same methods as, or be reduced to, the more precise physical sciences. Many would agree, too, that we cannot expect to find strict psychophysical laws. If there is anything new in what I say on this topic, it is in the details of the reasons I give for saying that generalizations that combine psychological and physical predicates are not lawlike in the strong sense that wholly physical laws can be. What apparently arouses the most doubt and opposition is my attempt to combine the view that psychological concepts have an autonomy relative to the physical with a monistic ontology and a causal analysis of action.

I thought, then, that my *conclusions* (in contrast, perhaps, to my arguments) concerning the nature of cognitive psychology as a science were neither new nor apt to excite much debate. But hearing Professor Peters's generous and sensitive remarks, as well as listening to the discussion of my paper at the Canterbury conference, made me realize I had given the impression that I was making some sort of attack on psychology generally, or at least on its right to be called a science. That is certainly not what I intended, but I do see how things I wrote could bear that interpretation. So here I will briefly try to set matters straight.

First, let me re-emphasize the fact that my arguments are limited in application to branches of psychology that make essential reference to 'propositional attitudes' such as belief, desire, and memory, or use concepts logically tied to these, such as perception, learning, and action. (Some of these concepts may not always show intensionality, and in such cases are also exempt.)

Second, I made much of the fact that psychophysical generalizations must be treated as irreducibly statistical in character, in contrast to sciences where in principle exceptions can be taken care of by refinements couched in a homogeneous vocabulary. This is not a *reproach* to psychology, nor does it mean its predictions and explanations are in fact less precise than those of many other sciences. I assume that in application meteorology and geology, for example, are far less precise than much work on perception. The point is not the actual degree of looseness in psychology, but what guarantees that it can't be eliminated, namely the conceptually hermaphroditic character of its generalizations.

Third, I argued that the part of psychology with which I was concerned cannot be, or be incorporated in, a closed science. This is due to the irreducibility of psychological concepts, and to the fact that psychological events and states often have causes that have no natural psychological descriptions. I do not want to say that analogous remarks may not hold for some other sciences, for example biology. But I do not know how to show that the concepts of biology are nomologically irreducible to the concepts of physics. What sets apart certain psychological concepts—their intensionality—does not apply to the concepts of biology.

'Science' being the honorific word it is in some quarters, it would be meretricious to summarize these points by saying that psychology (the part with which we are concerned) is not a science; the conclusion is rather that psychology is set off from other sciences in an important and interesting way. The argument against the existence of strict psychophysical laws provides the key to psychology's uniqueness: the argument led from the necessarily holistic character of interpretations of propositional attitudes to the recognition of an irreducibly normative element in all attributions of attitude. In the formulation of hypotheses and the reading of evidence, there is no way psychology can avoid consideration of the nature of rationality, of coherence and consistency. At one end of the spectrum, logic and rational decision theory are psychological theories from which the obviously empirical has been drained. At the other end, there is some form of behaviourism better imagined than described from which all taint of the normative has been subtracted. Psychology, if it deals with propositional attitudes, hovers in between. This branch of the subject cannot be divorced from such questions as what constitutes a good argument, a valid inference, a rational plan, or a good reason for acting. These questions also belong to the traditional concerns of philosophy, which is my excuse for my title.

Reply to Attfield. Mr Attfield believes that what I have called anomalous monism is inconsistent; in his view, psychological events cannot be identical with physical events while being nomologically unrelated to them. His argument, if I follow it correctly, has this form:

Premise 1. Some psychological event, say Attfield's perceiving of a fly at t, is identical with a certain physical event, say the neurological change in Attfield at t that is P.

Premise 2. Since there is a causal law that connects physical events with their causes, there is a causal law that connects the change in Attfield at t that is P with some physical event that caused it (say the change at $t-1$ that is P').

Conclusion. There is a (pschophysical) causal law that connects Attfield's perceiving of a fly at t with its cause, the change at $t-1$ that is P'.

Attfield is right that anomalous monism is committed to the premises, and that it rejects the conclusion. But I would urge that the conclusion does not follow from the premises. Attfield's argument has the same basic semantic structure as an argument that would infer that Jones believes Scott wrote *Waverley* from the facts that Scott is the author of *Waverley* and Jones believes the author of *Waverley* wrote *Waverley*. This argument fails because after psychological verbs like 'believes' normal substitutivity of coreferring singular terms breaks down. The same must be said about the positions occupied by 'a' and 'b' in 'There is a law that connects events a and b.' We cannot conclude from the fact that there is a causal law that connects events a and b, and the fact that $a = c$, that there is a causal law that connects events c and b. The reason is that laws (and nomological explanations) do not deal directly (i.e. extensionally) with events, but with events as described in one way or another.

The point may be made without reference to non-extensional contexts. Suppose there is an event that is uniquely characterized both by 'The event that is F at t' and 'The event that is G at t', and that there is a law, 'Every event that is H is followed a minute later by exactly one event that is F.' It does not follow that there is a law that says, 'Every event that is H is followed a minute later by exactly one event that is G.' Here 'H' and 'F' may be thought of as physical predicates, and 'G' as a psychological predicate.

Reply to Holborrow. Where Mr Attfield thinks that the identity of psychological and physical events entails the existence of psychophysical laws, Mr Holborrow is worried that unless psychophysical laws are introduced, the dualism of psychological and physical concepts will entail a dualism of autonomous 'causal systems'. 'At the psychological level,' he writes, 'the event which we describe as an action is caused by a reason. At the level of physical phenomena, the same event differently described is caused by an entirely different set of factors.' But causal relations, in my view, hold between events however described. So one and the same event, whether described as an action or as a physical event, will have the *same* causes, whether these are described as reasons or as physical states or events. There is no dualism of 'causal factors', 'causal systems', or 'types of causation'. Nor is there any reason to suppose, from the dualism of descriptions, that there are two kinds of law. If *a* caused *b*, then *some* descriptions of *a* and *b* instantiate a strict causal law. But the law is never, if I am right, a psychophysical law, nor can it be purely psychological (since the mental does not constitute a closed system).

Reply to Solomon. We cannot make good sense of the idea that there are seriously different total conceptual schemes, or frames of reference, or that there may be radically 'incommensurate' languages (to use Whorf's word). So I have argued elsewhere,[5] and Professor Solomon understandably wonders how I can reconcile my rejection of conceptual relativity with the claim that psychological concepts are nomologically and otherwise irreducible to physical concepts. He does make me regret saying that in interpreting the verbal behaviour of others we must translate into our own scheme of concepts. Of course interpretation is essentially translation, and so there is no avoiding that. But if translation succeeds, we have shown there is no need to speak of two conceptual schemes, while if translation fails, there is no ground for speaking of two. If I am right then, there never can be a situation in which we can intelligibly compare or contrast divergent schemes, and in that case we do better not to say that there is one scheme, as if we understood what it would be like for there to be more.

A conceptual scheme, in the context of the above remarks, is supposed to correspond to a whole language; nothing, at any rate, can be left out that is needed to make sense of the rest. A concept, or set of concepts, may be irreducible and yet be essential to making sense of some, or all, of the rest. Indeed if a concept *C* is essential to making sense of some other concept *C'* in a set of concepts *S*, *C* must be primitive (indefinable) relative to the resources of *S*.

Psychological concepts, I have been arguing, cannot be reduced, even nomologically, to others. But they are essential to our under-

[5] In 'The Very Idea of a Conceptual Scheme'.

standing of the rest. We cannot conceive a language without psychological terms or expressions—there would be no way to translate it into our own language. Of course there could be a part or fragment of a language that lacked psychological expressions, provided there was a (complete) language in which to incorporate or explain the fragment. It makes sense to speak of irreducible or semi-autonomous systems of concepts, or schemes of description and explanation, but only as these are less than the whole of what is available for understanding and communication.

13 *The Material Mind*

I want to discuss some general methodological questions about the nature of psychology as a science by assuming we know very much more than we do about the brain and the nervous system of man. Suppose that we understand what goes on in the brain perfectly, in the sense that we can describe each detail in purely physical terms—that even the electrical and chemical processes, and certainly the neurological ones, have been reduced to physics. And suppose, further, that we see that because of the way the system is constructed, the indeterminacies of quantum physics are irrelevant to our ability to predict and explain the events that are connected with input from sensation or output in the form of motion of the body.

While we are dreaming, let us also dream that the brain, and associated nervous system, have come to be understood as operating much like a computer. We actually come to appreciate what goes on so well that we can build a machine that, when exposed to the lights and sounds of the world, mimics the motions of a man. None of this is absurd, however unlikely or discredited by empirical discoveries it may be.

Finally, partly for fun, and partly to stave off questions not germane to the theme, let us imagine *l'homme machine* has actually been built, in the shape of a man, and out of the very stuff of a man, all synthesized from a few dollars' worth of water and other easily obtainable materials. Our evidence that we have built him right is twofold. First, everything we can learn about the physical structure and workings of actual human brains and bodies has been replicated. Second, Art (as I shall call him or it) has acted in all observable ways like a man: Art has had or seems to have had, appropriate expressions on his or its face, has answered questions (as it seems), and has initiated motions of a human sort when exposed to environmental change. Every correlation that has been discovered between what we know of mental processes, so far as this knowledge is reflected in physically describable ways, and what goes on in the human nervous system, every such correlation is faithfully preserved in Art. No one who did not know that Art was artificial would have discovered it by watching or listening, by prodding or talking. True, his makers could tell the observer exactly what was going on inside Art in terms of physics, and could explain in physical terms why Art moved as he did when subjected to various stimuli. But this should not put the observer on to the fact that Art came from the mad scientist's laboratory, since a similar explanation is possible in theory for men produced by more old-fashioned methods.

(The assumption that biology and neurophysiology are reducible to physics is not essential to the argument, and is probably false. Nor does anything really depend on the assumption that indeterminacy is irrelevant. Both assumptions could be eliminated, but at the expense of complicating the argument.)

And now the question is, what would all this knowledge of physics (and a fortiori of neurophysiology) tell us about psychology? Much less than might be expected, I shall argue, at least as long as we maintain a certain view of the subject matter of psychology.

For the scope of this paper, I shall treat psychology as a subject that deals with phenomena described by concepts that involve intention, belief, and conative attitudes like desire. I would include among these concepts action, decision, memory, perception, learning, wanting, attending, noticing, and many others. Attempts have been made, of course, to show that psychology can do without some or all of these concepts, for example by trying to define concepts like belief or desire in terms of concepts more behavioural, or otherwise more like the concepts used in the physical sciences. Direct elimination through definition of psychological terms no longer seems very plausible, and indeed if the line of argument I shall give is correct, definitional reduction is impossible. But of course other forms of reduction are imaginable. This fact marks a limit of the discussion: to the extent that psychology does not make essential use of the concepts I have described, the considerations that follow do not apply to it.

In any case, it would be foolish to maintain that the existence of Art would make no difference to psychology. It or he would show, for example, that determinism (to the extent that physics is deterministic) was compatible with every appearance of intentional action: aside from the matter of provenance, we would have as much reason to consider Art a voluntary agent as anyone else. Art would be as free as any of us, at least as far as could be told. And Art would prove that however different they may be, there is no conflict between the modes of explanation of physical science and of psychology.

Beyond these very general methodological questions, the existence of Art would no doubt have an influence on the direction and focus of research in the social sciences, on the design of experiments, and on the hypotheses considered worth testing. I take for granted that detailed knowledge of the neurophysiology of the brain will make a difference—in the long run, an enormous difference—to the study of such subjects as perception, memory, dreaming and perhaps of inference. But it is one thing for developments in one field to affect changes in a related field, and another thing for knowledge gained in one area to constitute knowledge of another. In a broad sense of relevance, I do not, of course, doubt the relevance of biology and the neurosciences to psychology. What interests me is that there seem to be limits to what can be directly learned from the other sciences (or from Art, as I am supposing) about psychology, and it is these limits I wish to explore.

It is time to be a little clearer about what did and what did not go into the manufacture of Art. Art is physically indistinguishable inside and out from a man, and he has reacted to changes in his environment by moving in ways indistinguishable from human behaviour. Identifiable parts of the interior of Art are physically connected with his movements, in accord with everything known about the construction of the brain and the nervous system. All this falls short, however, of assuming that we have succeeded in identifying such things as beliefs, desires, intentions, hopes, inferences, or decisions with particular states of the brain or mechanisms in it. Of course, there may be reason to connect *parts* of the brain with various cognitive processes; but parts are not mechanisms. And nothing in our description of Art requires that we be able to identify specific physical mechanisms with particular cognitive states and events. Since such states and events as thinking, believing, perceiving, and intending are conceptually central to all the psychological concepts (as I have arbitrarily designated them), we so far seem justified in saying that Art may not, directly at least, teach us much about psychology.

But how can this be? On a particular occasion, a pin penetrates the skin or surface of Art; he jumps away, wears the expression of pain and surprise, makes sounds like 'Ouch!' Or so we are tempted to describe matters. I assume we can describe the penetration of the skin, and all of Art's motions in purely physical terms—terms that can be incorporated into physical laws. Knowing the relevant structure of Art, we know *exactly* how the penetration of the skin caused the reaction (physically described). We also can describe cause and effect in more mundane ways—I just have. Now consider one pair of descriptions: the official physical description of the cause (or stimulus) and the psychological description of the effect (bodily movement, exclamation, facial expressions of surprise and pain). These are, we have agreed, descriptions of cause and effect and as such the events must fall under laws. If something like this holds for all psychological events—and we have been assuming nothing less —then are we not committed to the view that all psychological events are strictly predictable, and even that, for Art, we know how to predict them? Further, since we know both the physical and the psychological descriptions of the same events, why can we not correlate physical with psychological descriptions systematically? How then can we deny that in building Art we have reduced psychology to physics, and hence solved all the problems specific to psychology?

I would agree that we are committed to one important philosophical, and, indeed metaphysical, thesis. If psychological events cause and are caused by physical events (and surely this is the case) and if causal relations between events entail the existence of laws connecting those events, and these laws are, as we have supposed in designing Art, physical, then it must follow that psychological events simply *are* (in the sense of *are identical with*) physical events. If this is materialism, we are committed to it in assuming the existence of Art.[1]

Our commitments are less than might seem, however, for if I am right, we are not committed to the view that psychological events

[1] See Essay 11.

are predictable in the way physical events are; nor that psychological events can be reduced to physical events; nor that we have, in building Art, shown that we can explain psychological events as we can physical events. For I have not assumed, nor does anything I have assumed entail, that we can effectively correlate important open classes of events described in physical terms with classes of events described in psychological terms.

What I have supposed is that for any particular, dated psychological event we can give a description in purely physical terms; and so for any *given, finite* class of events, we can set up a correlation between psychological and physical descriptions. But although this can be done, it does not follow that such psychological predicates as '*x* desires his neighbour's wife', or '*x* wants a kaffee mit schlag', or '*x* believes that Beethoven died in Vienna', or '*x* signed a cheque for $20' which determine, if not infinite classes, at least potentially infinite ones;—it does not imply that such predicates have any nomologically corresponding physical predicates. Of course, if a certain class of psychological events is finite, and each psychological event has, as we are assuming, a physical description, then it follows trivially that there is a physical predicate that determines the same class as each psychological predicate. But this fact is in itself of no interest to science. Science is interested in nomological connections, connections that are supported by instances, whether or not the instances happen to exhaust the cases.

It should be easy to appreciate the fact that although every psychological event and state has a physical description, this gives us no reason to hope that any physical predicate, no matter how complex, has the same extension as a given psychological predicate—much less that there is a physical predicate related in a lawlike way to the given psychological predicate. To take an example from a different field that I have used before: consider some fairly rich language L that has the resources for describing any sentence of L. Assume in particular that L can pick out with a unique description each of the true sentences of L. But L cannot contain a predicate, no matter how complex, that applies to the true and only the true sentences of L—at least not if it is consistent. This fact would surprise someone who did not know about the semantic paradoxes. 'Surely,' he would say, 'since I can pick out each true sentence I can specify the class.' And he starts going through the true sentences, noticing what properties they have in common that none of the false sentences have. But he would be wrong; we know in advance that he cannot succeed. I think this is roughly the situation with psychological predicates in relation to physical: we know in advance that all the resources of physics will not suffice to pick out important (open or infinite) classes of events which are defined by psychological predicates.

We see, then, that complete knowledge of the physics of man, even if this covers, under its mode of description, all that happens, does not necessarily yield knowledge of psychology (a point made long ago by Plato's Socrates). Still, why should it not happen that there are inductively established correlations between physical and psychological events? Indeed, do we not already know that there are? We do, if by laws we mean statistical generalizations. The burned child avoids the flame (and psychology may contain more sophisticated examples). But these generalizations, unlike those of physics, cannot be sharpened without limit, cannot be turned into the strict laws of a science closed within its area of application. In giving my reasons for this conclusion, let me turn back again for a moment to the question what makes us think Art has been properly constructed from a psychological point of view. I think the answer has to be, Art gives every appearance of thinking, acting, feeling like a man. And not just the superficial appearances. If you cut him he bleeds, if you shine lights in his eyes, he blinks, and if you dissect his eyes, you discover rods and cones. It is important, in deciding that he has psychological traits, that he is made like a man. If we found a radio receiver inside, and learned that another person was sending out signals to make Art move, we would no longer be tempted to assign psychological characteristics to Art. Any important difference under the skin might make us hesitate. Nevertheless, our detailed understanding of the physical workings cannot, in itself, force us to conclude that Art is angry, or that he believes Beethoven died in Vienna. In order to decide this, we would have first to observe Art's macroscopic movements, and decide how to interpret them, in just the way we decide for humans.

It would be easy to go wrong in our thinking here, partly because I have assumed we deliberately *built* Art to do what he does. And probably in building Art, we used circuits of the sort we would use if we wanted to build a machine that could process information, and so forth. But of course we must not jump to the conclusion that when those circuits come into play, Art is processing information.

The Material Mind 251

Much of what is at stake is whether what would be information for us, if Art were merely an extension of our own faculties (as a computer is), is information for him. To assume this point is to suppose Art sees things as we do, and means by the sounds he makes what we should mean. But this we can decide only by seeing how such assumptions fit into the total picture of Art's behaviour. The point is a simple one. If we want to decide whether Art has psychological properties, we must stop thinking of him as a machine we have built and start judging him as we would a man. Only in this way can we study the question of possible correlations between physical and psychological properties.

It will be best to admit at this point that the fact that Art is artificial plays no essential part in the argument. The reason is that I have not supposed that he was built on the basis of knowledge of *laws* correlating psychological and physical phenomena: all that was known was the physical correlate of each *particular* movement or act. It is true that we can predict Art's physical movements. But if we want to know whether a particular one of these will be interpretable as an action or response, we can tell only by considering all the physical aspects in detail (including of course what the environment will be like) and then judging the case as we would a human movement. We have no clear reason to say that Art will continue to seem human. So Art proves no point that cannot be made as well by supposing we have the same sort of comprehensive knowledge of the physics of a man as we have pretended we have of Art. Art served the heuristic purpose of removing any mysterious unknown properties. But in fact all we removed was unknown *physical* properties, and we can suppose these removed from a man as easily as from Art. The supposition no more settles the question whether man has a soul (i.e., irreducible psychological properties) than it settles the question whether we *gave* Art a soul.

I return again to the question why we should not expect to discover sharp lawlike correlations (or causal laws) connecting psychological and physical events and states—why, in other words, complete understanding of the workings of body and brain would not constitute knowledge of thought and action. Before I give what I think is the correct reason, let me briefly mention some bad reasons that have commonly been offered.

It is often said, especially in recent philosophical literature, that there cannot be a physical predicate with the extension of a verb of action (for example) because there are so many different ways in which an action may be performed. Thus a man may greet a woman by bowing, by saying any of a number of things, by winking, by whistling; and each of these things may in turn be done in endless ways. The point is fatuous. The particulars that fall under a predicate always differ in endless ways, as long as there are at least two particulars. If the argument were a good one, we could show that acquiring a positive charge is not a physical event, since there are endless ways in which this may happen.

There is a symmetrical argument that is equally common and equally bad: it is said that the same physical event may count as quite different actions. So, for example, exactly the same motion and sound emanating from an agent may on one occasion constitute a greeting and on another occasion constitute an insult. But of course if the occasions differ, the events must differ in some physical characteristics. The difference may lie within the agent. There may, for example, be a difference in intention: this difference, we assume, has its physical aspect, since it is reflected in the propensities to motion of the agent. Given a complete description of the brain, we must expect this difference to correspond to some difference in physiology—ultimately in physics, as we have been seeing it.

We can imagine cases, however, where the intention is the same, and the beliefs and desires, and so also everything physical in the agent; and yet different actions are performed. Thus a man might intend to keep a promise by going to the opera. Yet on one occasion his going to the opera with this intention might constitute the keeping of a promise, and on another occasion not (he might have forgotten the day). But again the physical situation is not identical in all physical respects. We simply must define the physical event or situation more broadly—just as keeping a promise depends on certain prior events having taken place, so the occurrence of a physical event of a certain sort may depend on a broad physical setting in which it takes place. If we pleased, we could define a supereclipse of the moon to be an eclipse that was preceded, within two weeks, by an eclipse of the sun. A supereclipse may not be of much interest to science, but it is surely a respectable physical concept.

Again, it is said that cultural relativism affects the classification of actions, but not of physical events. So the same gesture may indicate assent in Austria and dissent in Greece. Here we need only increase the frame of physical reference to find a relevant difference: Austria is physically distinct from Greece, and so any event in Austria is physically distinct from any event in Greece. Perhaps it will be suggested that the same *particular* gesture of a man may be judged to be an act of assent by an Austrian and an act of dissent by a displaced Greek. In this case, however, the two descriptions cannot contradict one another. Just as an object may accelerate relative to one frame of reference and not relative to another, so a gesture may count as assent *to an Austrian* and as dissent *to a Greek*. Only if we accept an unduly restricted view of the predicates that can be formed using physical concepts are we apt to be attracted by any of these arguments.

In these considerations, two important themes emerge. One is the necessity for distinguishing individual, dated, events from sorts of events. We may without error say that 'the same gesture' has one meaning in Austria and another in Greece: what we have in mind, of course, are gestures of *some relevant same sort*.[2] The other theme concerns the relation between psychological descriptions and characterizations of events, and physical (or biological or physiolog- ical) descriptions. Although, as I am urging, psychological charac- teristics cannot be reduced to the others, nevertheless they may be (and I think are) strongly dependent on them. Indeed, there is a sense in which the physical characteristics of an event (or object or state) *determine* the psychological characteristics; in G. E. Moore's view, psychological concepts are *supervenient* on physical concepts. Moore's way of explaining this relation (which he maintained held between evaluative and descriptive characteristics) is this: it is impossible for two events (objects, states) to agree in all their physical characteristics (or in Moore's case, their descriptive characteristics) and to differ in their psychological characteristics (evaluative).

The two themes, of the distinction between individual events and sorts, and the supervenience of the psychological on the physical, are related. For what needs to be stressed is that it is the descriptions of individual psychological events, not sorts of events, that are supervenient on physical descriptions. If a certain psychological concept applies to one event and not to another, there must be a difference describable in physical terms. But it does not follow that

[2] See Essays 9 and 10.

there is a single physically describable difference that distinguishes any two events that differ in a given psychological respect.

There is another class of arguments that I cannot deal with in any detail: these are arguments based on the claim that psychological concepts are essentially evaluative, while physical concepts are not. If this means that when we call an event an action, we are not, or not merely, describing it, but are also judging it as good or bad, blameworthy or reasonable, then I think this is wrong. Whenever we say anything, we may be expressing a value of some sort; but this does not mean that what we say may not also be true or false. In any case, to make sense of the question why there are no strict laws connecting physical and psychological phenomena, we must assume that judgements concerning these phenomena are true or false in the same way.

In a quite different sense, evaluative considerations may be thought to enter our judgements about the actions people perform. It may be claimed that there are certain regulative or constitutive elements in the application of psychological concepts. This is certainly right; but the same can be said for the application of physical concepts. Nevertheless, here we are much closer to the truth.

Let us consider a particular historical event, say David Hume's admitting in an appendix to his *Treatise* that he cannot see how to reconcile two of his theses. Making an admission is necessarily an intentional act, and it entails that what is admitted is the case—in our example, Hume's admission entails that he cannot see how to reconcile the two theses. Since making the admission was intentional, we also know that Hume must have believed that he did not see how to reconcile the two theses, and he must have wanted (probably for some further reason) to reveal this fact. Not only did Hume have this desire and this belief, but they were somehow efficacious in his making the admission—he made the admission because he had the desire and the belief. If we interpret this 'because' as implying (among other things) a causal relation—and I believe we must—then in describing an action as performed with a certain intention, we have described it as an action with a certain causal history. So in identifying the action with a physical event, we must at the same time be sure that the causal history of the physical event includes events or states identical with the desires and cognitive states that yield a psychological explanation of the action.

This is only the beginning of the complications, however, for most

The Material Mind   255

emotional states, wants, perceivings, and so on, have causal connections with further psychological states and events, or at least require that these other states exist. And so, in saying an agent performed a single intentional action, we attribute a very complex system of states and events to him; all this must be captured in giving the corresponding physical states and events. I am not, of course, arguing that there is not a corresponding physical description—I am sure there is. I am not even arguing that we could not produce the corresponding description in particular cases. I am trying to show only why we cannot establish general, precise, and lawlike correlations between physical and psychological descriptions. The complexity of psychological attributions does not in itself prove the point. But it will turn out that the quality of this complexity is germane.

At this point it will help to turn to a psychological phenomenon one step more abstract—the ability to speak and understand language. We cannot hope in any case to cope with the full range and subtlety of psychological traits without taking account of language, for the finer distinctions among desires and beliefs, thoughts and fears, intentions and inferences, depend on the assumption of a cognitive structure as complex as that of language and cannot be understood apart from it.

In the end, we want to be able to explain speech acts, which are intentional and have the characteristics of other actions recently touched on. Part of explaining such acts is interpreting them, in the sense of being able to say what the speaker's words expressed on an occasion of use—expressed in his language, of course. We have a full grasp of what a man said when he uttered certain sounds only if we know his language, that is, are prepared to interpret a large number of things he might say. For we do not understand a particular sentence uttered by a man unless we know the role the words in it play in other sentences he might utter. To interpret a single speech act, therefore, we must have a grasp of the speaker's unrealized dispositions to perform other speech acts. Indeed, we may think of having, or knowing, a language as a single, highly structured, and very complex, disposition of the speaker. We describe the disposition by specifying what the speaker would mean by uttering any of a large number of sentences under specified conditions.

Described psychologically, a speaker's language ability is a complex disposition. Described physically, it is not a disposition, but an actual state, a mechanism. So here, if anywhere, it would seem that detailed knowledge of the physical mechanism should be a help to psychology. No doubt in each man there is some physical state, largely centered in the brain, that constitutes his language ability. But how can we identify this state? (I do not mean merely locate it, but describe in detail the relevant mechanism.) How do we know that a certain physical state of the brain, a certain mechanism, is the mechanism that accounts for the speaker's speech behaviour, his saying and meaning what he does when he speaks? I assume, as before, that if the agent speaks, we can on each occasion identify the particular physical event that corresponds. Thus there is no problem about testing the claim that a particular physical mechanism (for example Art) is a language-speaking mechanism: we can test it just as we test a man for language ability, by noticing how it behaves in various circumstances. This will not, however, give us what we want, which is a lawlike correlation between workings of the mechanism and speech behaviour. We want to know what the physical property of the machine—of any machine—is that would make it speak like a man.

Why can we not simply say: the physical property is just the one that produces the observed results? This is inadequate, because the required results outrun the observed ones: we want the physical property that would produce linguistic behaviour. Here we do have one description of the physical property, but it is a description that uses psychological concepts. It is like saying man is a language-speaking machine. True; but what does the word 'machine' tell us?

We interpret a single speech act against the background of a theory of the speaker's language. Such a theory tells us (at least) the truth conditions of each of an infinite number of sentences the man might utter, these conditions being relative to the time and circumstances of utterance. In building up such a theory, whether consciously, like an anthropologist or linguist, or unwittingly, like a child learning its first language, we are never in a position directly to learn the meanings of words one by one, and then independently to learn rules for assembling them into meaningful wholes. We start rather with the wholes, and infer (or contrive) an underlying structure. Meaning is the operative aspect of this structure. Since the structure is inferred, from the point of view anyway of what is needed and known for communication, we must view meaning itself as a theoretical construction. Like any construct, it is arbitrary except for the formal and empirical constraints we impose on it. In the case of meaning, the constraints cannot uniquely fix the theory of interpretation. The reason, as Quine has convincingly argued, is that the sentences a speaker holds to be true are determined, in ways we can only partly disentangle, by what the speaker means by his words and what he believes about the world. A better way to put this would be to say: belief and meaning cannot be uniquely reconstructed from speech behaviour. The remaining indeterminacy should not be judged as a failure of interpretation, but rather as a logical consequence of the nature of theories of meaning (just as it is not a sign of some failure in our ability to measure temperature that the choice of an origin and a unit is arbitrary).

Underlying the indeterminacy of interpretation is a commonplace fact about interpretation. Suppose someone says, 'That's a shooting star.' Should I take him to mean it really is a star, but that he believes some stars are very small and cold; or should I think he means it is not a star but a meteorite, and believes stars are always very large and hot? Additional evidence may resolve this case, but there will always be cases where all possible evidence leaves open a choice between attributing to a speaker a standard meaning and an idiosyncratic pattern of belief, or a deviant meaning and a sober opinion. If a speaker utters the words, 'There's a whale', how do I know what he means? Suppose there is an object that looks like a whale in the offing, but I know it is not a mammal? There seems to be no absolutely definite set of criteria that determine that something is a whale. Fortunately for the possibility of communication, there is no need to force a decision. Having a language and knowing a good deal about the world are only partially separable attainments, but interpretation can proceed because we can accept any of a number of theories of what a man means, provided we make compensating adjustments in the beliefs we attribute to him. What is clear, however, is that such theory construction must be holistic: we cannot decide how to interpret a speaker's 'There's a whale' independently of how we interpret his 'There's a mammal', and words connected with these, without end. We must interpret the whole pattern.

At this point, we might hope that knowledge of the physical correlate of the speech mechanism would be a help. After all, words are used as they are because of the way this mechanism works. Can we not locate *the physical correlates of meaning*? Can we not discover unambiguously on the physical level what we must merely infer, or treat as a construct, as long as we stick to observation of speech behaviour?

Well, how might it work? We might find out exactly what patterns of sights and sounds and smells, described now in terms of physical inputs, suffice to dispose our artful machine to utter, 'That's a whale', when asked, 'What is that?' (And so for endless further cases.) Would we then know what Art means? I think the answer is, we would know no more and no less about meaning than we do about human speakers now. For what would Art say if he 'learned' that an object with a cetateous appearance was not a mammal? How can we decide without knowing what he means by 'mammal'? Suppose the whale were to appear very small, or upside down, but that Art 'believes' he is looking through the wrong end of a telescope, or inverting glasses? A few questions like this should make us realize that we cannot simply associate some fixed part of Art's brain, or aspect of it, with the criteria for the application of a word.

Might we not identify the meaning of a sentence with the intention with which it is spoken, and hunt for the physical correlate of the intention, thus avoiding the problem of endless ramifications that seems to plague theories of meaning or interpretation? The difficulty is that specific intentions are just as hard to interpret as utterances. Indeed, our best route to the detailed identification of intentions and beliefs is by way of a theory of language behaviour. It makes no sense to suppose we can first intuit all of a person's intentions and beliefs and then get at what he means by what he says. Rather we refine our theory of each in the light of the other.

If I am right, then, detailed knowledge of the physics or physiology of the brain, indeed of the whole of man, would not provide a shortcut to the kind of interpretation required for the application of sophisticated psychological concepts. It would be no easier to interpret what *l'homme machine* means by what it 'says' than to interpret the words of a man, nor would the problem be essentially different. (There would be one unimportant shortcut: where with a man we must gather our evidence by creating experimental situations, we could disassemble the machine. But after disassembly, we could only say, in psychological terms, what the machine would do under completely specified circumstances; no general laws about its behaviour would be forthcoming.) With the machine, then, as with the man, we would have to interpret the total pattern of its observed (or predicted) behaviour. Our standards for accepting a system of interpretation would also have to be the same: we would have to make allowance for intelligible error; we would have to impute a large degree of consistency, on pain of not making sense of what was said or done; we would have to assume a pattern of beliefs and motives which agreed with our own to a degree sufficient to build a base for understanding and interpreting disagreements. These conditions, which include criteria of consistency and rationality, can no doubt be sharpened and made more objective. But I see no reason to think that they can be stated in a purely physical vocabulary.

Past discoveries about the nature of the brain, and even more, the discoveries we can expect from workers in this field, throw a flood of light on human perception, learning, and behaviour. But with respect to the higher cognitive functions, the illumination must, if I am right, be indirect. There is no important sense in which psychology can be reduced to the physical sciences.

14 *Hempel on Explaining Action*

In December of 1961 Hempel gave the presidential address at the annual meeting of the Eastern Division of the American Philosophical Association. The title was 'Rational Action'.[1] In that address, Hempel argued that explanation of intentional action by appeal to the agent's reasons does not differ in its general logical character from explanation generally; in taking this position, he was swimming against a very strong neo-Wittgensteinian current of small red books. Two years after Hempel read his address to the A.P.A. I published 'Actions, Reasons and Causes';[2] it made some closely related points. Rereading these two papers in tranquillity, two things strike me. The first is how subtle and careful Hempel's paper is, how well the arguments and claims in it stand up after the subsequent years of discussion and controversy, and with what imagination and skill Hempel foresaw and tried to answer the difficulties that would arise in a sceptical reader's mind. The second thing that strikes me is that my paper made no reference to Hempel's. My excuse is perhaps that Hempel put his case so persuasively and at the same time so modestly that I found it easy to assume I was born knowing it all. (It is also possible I didn't get to read Hempel's piece until later. However, I also didn't mention Ducasse, whose 1926 article 'Explanation, Mechanism, and Teleology'[3] anticipated us both, and which I *know* I had read, since it was in Feigl and Sellars, and I got through graduate school by reading Feigl and Sellars.) Anyway, I cannot say how glad I am for this opportunity to make amends. My debt to Peter, intellectual and otherwise, is more

[1] Carl G. Hempel, 'Rational Action'.
[2] Essay 2.
[3] C. J. Ducasse, 'Explanation, Mechanism, and Teleology'.

than I can count. It started accumulating when we were colleagues almost thirty years ago, and he gently and generously helped me with my first faltering lectures in elementary logic; and I know the debt will grow today; and far into the future.

On one issue, my early paper and Hempel's earlier one did differ. I emphasized the role of *causality* in our understanding of action, urging that an appropriate belief and desire could explain, and be the reasons for, an action only if they caused it. I don't think Hempel objects to this idea, and indeed in the expanded version of his A.P.A. address which appeared as part of the final essay in his 1965 book he explicitly says that 'the offer of a bribe . . . may be said, in everyday parlance, to have caused the explanandum event [treason]'.[4] The difference in our accounts, if there really is one, concerns the exact way in which *laws* are involved when we explain actions by mentioning the agent's reasons.

There is a weak sense in which laws may be said to be involved which is not in dispute. Hempel holds, and I agree, that if *A* causes *B*, there must be descriptions of *A* and *B* which show that *A* and *B* fall under a law. This is a weak thesis, because if this is the only sense in which laws must be invoked in reason explanations, someone might explain an action by giving the agent's reasons while having no idea what the relevant law was. I have argued that a causal relation implies the existence of strict laws belonging to a closed system of laws and ways of describing events, and that there are no such laws governing the occurrence of events described in psychological terms; we seldom if ever know how to describe actions or their psychological causes in such a way as to allow them to fall under strict laws.[5] It would follow that we can explain actions by reference to reasons without knowing laws that link them.

Hempel, if I am right, believes all explanation requires reference, oblique or direct, to relevant known empirical laws. In that case, in order to explain events we must describe them in a way that reveals how laws are applicable.

This sounds like a forthright conflict, but in fact I'm not sure it is. For on the one hand, I'm not certain what Hempel requires us to know about the relevant law; and on the other hand, I haven't denied that there may be laws far less than strict or deterministic that we must know or assume to be true when we explain actions.

[4] Carl G. Hempel, *Aspects of Scientific Explanation*. [Hereafter referred to as *Aspects*.]   [5] Essay 11.

We may join in lauding as an ideal explanation a description of antecedents and a specification of laws such that the explanandum can be deduced; but how much less still counts as explanation? It seems to me we have in action a particularly good specimen for study; since we agree that one way of explaining actions is by giving the agent's reasons, we can concentrate on the relatively clear question what reason explanations are like, and set aside the more diffuse problem of characterizing explanation generally.

Let me give a simple—and obviously false—account of how the reasons for an action explain it. I think any correct account is a modified version of it. Ada presses in succession the seven, the two, the nine, and the cube root buttons on her calculator. Her reasons are that she wants to take the cube root of 729 and she believes that by pressing in succession the seven, two, nine and cube root buttons she will take the cube root of 729. We may imagine her reasoning as follows: any action of mine that is a taking of the cube root of 729 is desirable; my pressing in succession the seven, two, nine, and cube root buttons will be such an action; so, my pressing those buttons is desirable. Parallel to her little piece of reasoning is ours: Ada wants to take the cube root of 729. A want is, or entails, a certain disposition to act to obtain what one wants. That someone has a certain disposition may be expressed as a generalization or law governing the behaviour of that person. In the present case, if Ada thinks some action will constitute her taking the cube root of 729, she will take that action. She thinks that pressing the 7, 2, 9, and cube root buttons in succession will constitute such an action. So, she presses the buttons. The parallel between Ada's reasoning, and our explanation, gives a fairly precise meaning to Aristotle's claim that the conclusion of a piece of practical reasoning is an action.

It is obvious that in this cartoon, our description of Ada's reasons provides a classical nomological–deductive explanation of her action. Our description of her desire expands into a 'law', while the description of her belief provides the singular premise which triggers the wanted conclusion.

One trouble is, of course, that the 'law' is false. Far more often than not people fail to perform any action at all to achieve a desired end, even though they believe or know the means are at hand; and no one ever performs all the actions he believes will lead to the end (unless there is nothing he believes can be done, or for some reason only one action is thought possible). But we can say of someone who has a desire or end that he will *tend* to behave in certain ways under specified circumstances.[6] If we were to guess at the frequency with which people perform actions for which they have reasons (not necessarily adequate or good reasons, but reasons in the simple sense under consideration), I think it would be vanishingly small. (To aid your imagination: what is the ratio of actual adulteries to the adulteries which the Bible says are committed in the heart?)

While the probability may be very low, it isn't zero; and on the other side, we can say that it is (logically) impossible to perform an intentional action without some appropriate reason. So a desire (represented now by a probabilistic law) and a belief do provide at least a very low-grade statistical explanation. If the agent had the appropriate belief and desire, it was at least possible for him to perform the action to be explained, and there was some (very low) probability that he would.

This does not seem to me to give a very convincing account, however, of why we count reason explanations of actions as so meaningful—as such good explanations. It may do as well as we can do to explain why a coin comes up heads four times in a row to say, well, that does happen on average once every so often. We feel we do very much better in explaining why Ada punched those keys when we mention her desire and belief—far better than could be accounted for by the fact that once in a blue moon someone with those motives performs such an act.

The reason the explanation is so good, it seems to me, is that motives explain an action only if they cause it, and in a very special way. A person may have certain motives for an act, and yet perform it either by accident or for quite different reasons. So reasons explain an action only if the reasons are efficacious in the situation. And even this is not enough; a man's motives for acting in a certain way may cause him to act in that way without it's being the case that those were his reasons for performing the act. Thus a man might want to break a pot, and believe that by stamping on the floor he will cause the pot to break. The belief and desire cause him to stamp, but the stamping has no direct effect on the pot. However, the noise makes a bystander utter an oath which so offends the agent that he swings around, accidentally knocking over and breaking the pot. The agent had a motive for breaking the pot, and the motive caused

[6] *Aspects*, 473. Compare Essay 2.

him to break the pot. But he had no motive in breaking the pot: it was an accident.[7]

That a complex causal story goes with a reason explanation of action doesn't show that laws aren't needed; if Hempel is right,

Causal explanation is a special type of deductive nomological explanation; for a certain event or set of events can be said to have caused a specified 'effect' only if there are general laws connecting the former with the latter in such a way that, given a description of the antecedent events, the occurrence of the effect can be deduced with the help of laws.[8]

We can't tell from this quotation whether a singular causal claim ('*A* caused *B*') explains the occurrence of *B* provided a true law connects *A* and *B* under some descriptions of *A* and *B*; or whether the law must cope with the very expressions '*A*' and '*B*'. Further passages in Hempel strongly suggest the latter:[9] the descriptions of cause and effect must at least come close to giving the form of the required law. Applied to action explanations, I think this means that we must have some idea how to turn the very rough low-grade statistical tendency statements implied by attributions of desires and beliefs into serious empirical laws. How might this be done? The method at which Hempel hints is that we should refine the law by making a fuller statement of the conditions under which the agent will act. And we can see in a general way how this is often done.

Before going on, I need to bring out what may be an important difference between the line I have been developing and Hempel's views as expressed in his A.P.A. address and *Aspects of Scientific Explanation*. The line I have been developing suggests that the laws implicit in reason explanations are simply the generalizations implied by attributions of dispositions. But then the 'laws' are peculiar to individuals, and even to individuals at particular moments. So if a vagrant fancy for a ride on the fun wheel were to flit through Dora's mind, some law like this would be true: for the space of those few moments, if Dora had believed some action of hers would have produced a ride on the fun wheel (and a lot more conditions were satisfied), Dora would have taken that action. If this is the sort of law involved in reason explanations, we all know an enormous number of highly particular laws. It does not sound like much of an argument for the possibility of a scientific psychology.

[7] A similar point is made at the end of Essay 4.
[8] *Aspects*, 300–1. [9] e.g. *Aspects*, 348–9.

Reflections of this sort may explain why Hempel takes a somewhat different tack. His idea, if I am right, is to introduce laws of far greater generality which say something about what all agents will do under certain conditions. At one point, for example, he says that references to motivating reasons 'provide explanatory grounds for the resulting actions only on the assumption that people motivated by such and such reasons will *generally* act, or will *tend* to act, in certain characteristic ways'.[10] Thus a more general law governing Dora's case might be: If *anyone* wants to take a ride on a fun wheel, then (given further conditions) he or she will do what he or she believes will result in a ride. Particular premises will then apply this law to Dora: she does want to take a ride, and believes that . . . etc.

The law may be made more general still: Hempel speaks of an explanation that 'clearly presupposes psychological assumptions about the manner in which an intelligent individual will tend to act, in the light of his factual beliefs, when he seeks to attain a given objective'.[11] In his A.P.A. address, Hempel puts the matter in terms of rationality. A reason explanation, when filled out, gives, in addition to the relevant beliefs and desires of the agent, a statement of what a rational agent will do given such beliefs and desires (and perhaps further conditions). The explanation is then completed by stating that the particular agent was rational.

There are difficulties, I think, with this idea. Hempel says rationality is a kind of character trait: some people have it and some don't, and it may come and go in the same individual. No doubt some people are more rational than others, and all of us have our bad moments. And perhaps we can propose some fairly objective criteria for testing when someone has the trait; if so, knowing whether someone is rational at a given time may help us to explain, and even predict, his behaviour, given his beliefs and desires. But reference to such a trait does not seem to me to provide the generality for reason explanations that Hempel wants. For in the sense in which rationality is a trait that comes and goes, it can't be an assumption needed for every reason explanation. People who don't have the trait are still agents, have reasons and motives, and act on them. Their reasons are no doubt bad ones. But until we can say what their reasons are—that is, explain or characterize their actions in terms of their motives—we are in no position to say the reasons

---

[10] *Aspects* 451.   [11] *Aspects* 450.

are bad. So being in a position to call a person rational, irrational, or nonrational in this sense presupposes that we have already found it possible to give reason explanations of his actions.

A trait of rationality is too blunt an instrument to help much in all reason explanations, for even the most rational person will often do things for poor reasons, while, as I have said, the looniest action has its reason. What is needed, if reason explanations are to be based on laws, is not a test of when a person is rational, but of when a person's reasons—his desires and beliefs—will result, in the right way, in an action. At this point the assumption of rationality seems in danger of losing empirical content. For now the explanation of Dora's action requires, in addition to the statement of the relevant desire and belief, a general law that says: Anyone who is rational with respect to certain beliefs and desires at a certain moment will act on them; Dora was rational with respect to such and such beliefs and desires at that moment. So, she boarded the fun wheel at that moment. It does not seem likely that we can give a testable content to the required relativized concept of rationality that would allow us to test this law empirically. But if we cannot, the so-called law merely states part of what we mean, in this context, by saying an action is rational. The only empirical generalization on which the explanation rests is given when we describe Dora's beliefs and desires.

The same point may be made with respect to the attempt at generalizing more particular relations between desires and actions. We considered, for example, the 'law' that says: If anyone wants to take a ride on a fun wheel, he or she will tend to do what he or she believes will result in a ride. It is hard to think what evidence we would accept as seriously discrediting this claim. The point is not that nothing but action (accompanied by appropriate beliefs) reveals a desire, but that nothing can definitely override it.

Hempel appreciates the danger that the assumption of rationality may lack empirical import, but he defends the assumption by pointing out that people often do fail to act on the basis of reasons they have, or reason badly, or ignore relevant information. This shows, he argues, that the assumption of rationality is not empty. It certainly is an empirical matter that people sometimes act on reasons they have and sometimes don't. But this does not show that we can state, in an objectively determinable way, when people will and when they won't act on their reasons. Unless we can do this, though, the assumption of rationality suggests no empirical law that could be used in reason explanations generally.

The discussion so far has been hampered, if not hamstrung, by my sticking to a particularly simple form of reason explanation, and this has prevented me from saying anything sensible about a number of problems, such as how an agent might be expected to choose among several competing actions, each of which is recommended by reasons he has. Similarly, no mention has been made of the effect of variations in the strength of desire, or degree of belief. The theory of decision making under uncertainty is designed to cope with these matters. Since it offers a more sophisticated way of dealing with reason explanations, and is still based on common sense ideas of how actions are explained, let me turn to it.

For our purposes, the theory may be considered as having two parts. The first part describes a rational pattern of preferences. Someone has a rational pattern of preferences among a set of alternatives provided his preferences meet a list of requirements. I won't try to state these requirements, but most of them are intuitively obvious: his preferences are transitive, asymmetric, and connected in the set of alternatives, and his preferences among alternatives that involve risk reflect in a systematic way the assumption that he values an alternative in proportion to how likely he believes that alternative is to produce valued outcomes. If a man has a rational pattern of preferences, it is possible to assign numbers to the alternatives that measure their relative values to him, and other numbers to events that register how probable he thinks it is that they will occur. The claim that a rational pattern of preferences *is* rational may be justified in this sense: if someone has a set of preferences that is not rational, it is possible to make book against him in such a way that whatever happens he will lose out by his own standards.

The second part of the theory relates action to preferences: it states that a person with a rational pattern of preferences always chooses an alternative (from among those available to him at the moment) such that no other has a higher expected value.

And now, the hypothesis that a particular person, or that all people, act rationally in the sense of decision theory would certainly seem to be empirical, testable, and probably false. So I assumed when I spent three years or so doing experiments designed to test the theory, and I'd like to describe some of what I think I learned. Fortunately for you, and unfortunately for me, I can put it briefly.

First a remark on the relations between decision theory and ordinary reason explanations. Pretty obviously particular choices can be explained in decision theory along lines very similar to those followed in reason explanations: a particular action is chosen from among the available set because of the agent's beliefs (for example how apt he thinks the action is to produce various results) and the relative values he sets on the possible outcomes. His desires are thus made comparative and quantitative. Some ordinary desires, however, do not translate directly into preferences, so that not all reason explanations have a clear decision theory counterpart explanation. This is particularly evident if we think of conflicting desires. A person may have a reason for preferring *A* to *B* and another reason for preferring *B* to *A*. This is an embarrassment for reason explanations, for they need to predict which reason will win out. Decision theory skips the problem, for it says nothing about why one basic outcome is preferred to another, and the theory bars possible evidence of conflict in behaviour. Neither approach suggests an interesting account of how conflict can arise or be resolved.

What I want to discuss is the empirical content of the hypothesis that people are rational in the sense of decision theory. I stumbled on the subject when I read an article by Mosteller and Nogee, 'An Experimental Measurement of Utility'.[12] They calculated the relative values of winning various amounts of money for each of their subjects, going on the assumption that the relevant probabilities were objectively known (e.g. that a normal coin has a probability of 1/2 of coming up heads). Discussing earlier work, they noted that other experimenters, using exactly the same sort of wagers, had calculated the subjective probabilities of various events (e.g. of a coin coming up heads) on the assumption that utility is linear in money. It did not seem to worry Mosteller and Nogee that the same experiment could be interpreted in such different ways. Part of the trouble came, perhaps, from overdependence on von Neumann and Morgenstern's axiomatization of utility in their 1944 book *Theory of Games and Economic Behaviour*,[13] which simply took probabilities as given. At the time no one seemed to have any idea how to proceed experimentally if neither probabilities nor utilities were known; yet there was no more reason to assume one to be known

[12] F. Mosteller and P. Nogee, 'An Experimental Measurement of Utility'.
[13] J. von Neumann and O. Morgenstern, *Theory of Games and Economic Behavior*.

independently than the other. In fact, however, the problem had been brilliantly solved by Frank Ramsey as early as 1926.[14] Ramsey put conditions on preferences between wagers while making no untestable assumptions about probabilities or utilities, and he outlined a method for testing the theory experimentally. Ramsey's work was not known to von Neumann and Morgenstern, nor to any of the experimentalists, nor, so far as I know, to economists who were then working on utility theory. But Mosteller and Nogee mentioned Ramsey, apparently without having read him, in a footnote; they had heard from Donald Williams that Ramsey had something to say on the subject. That led me to read Ramsey, and I realized at once that Ramsey had solved the problem of disentangling the roles of subjective probability and utility—of belief and desire—in decision making under uncertainty. Patrick Suppes, Sidney Siegel, and I then did a series of experiments designed to test whether people actually make choices as Ramsey says they should; whether, that is, people are rational to the extent that their utilities and degrees of belief can be measured in the appropriate ways.[15]

As time went on, I became more and more sceptical about what the experiments showed. Of course they showed *something*. All we had to do was to give a clear behaviouristic interpretation to 'S prefers A to B' and decision theory—that is, a certain set of sentences containing the previously uninterpreted word 'prefers' —became a powerful empirical theory, eminently testable, and palpably false. We were happy with an interpretation that made the theory look more or less true over a very limited set of alternatives. If things didn't quite work out, there was a wealth of possible explanations. For one thing, there was always the question when an alternative or event was 'the same' one as in an earlier trial. A subject, asked to choose between A and B, and then between B and C, might not regard B in the company of A as the same alternative as B in the company of C. Or, an event described as this coin coming up heads might not be viewed as having the same probability on a first throw as after five throws. Or perhaps the subject might not understand English: interpreting his choice behaviour depended on assuming he could interpret our speech behaviour.

In psychophysical experiments it is often assumed that the rela-

[14] F. P. Ramsey, 'Truth and Probability', in *The Foundations of Mathematics*.
[15] Donald Davidson, Patrick Suppes, and Sidney Siegel, *Decision Making: An Experimental Approach*.

tive frequency of yes to no answers to questions like, 'Is this sound louder than that?' can be used to induce a subjective interval scale on the objectively identified stimuli. The idea is that the further apart subjectively two stimuli are, the more often they will be discriminated. There seemed no reason not to use the same technique for scaling utilities. If the experimenter forces a choice, a subject will choose *A* half the time and *B* the other half if they are equal in value to him; the greater his preference for *A* over *B*, the more often he will choose *A*. This method gives a second method for scaling utilities, and this suggested to Jacob Marschak and me that we see whether the scale induced by this method was the same as the scale obtained by the Ramsey method involving risk. I won't describe the experiment, which was very complicated, nor the results, which were inscrutable.[16] But suppose the two methods had given different results; which one would be the test of utility? And what should one say anyway about all those pairs of options between which subjects vacillated? That the subject could not tell which one he preferred? Or that he changed his mind from time to time?

The last question began to bug me: how could we tell that subjects weren't influenced in their preferences by the experiment itself—that their preferences weren't changing as we went along? Decision theory purports to describe a static situation: the pattern of a person's attitudes and beliefs at a moment. Of course we tried hard not to change the preferences of a subject during a session. If you offer a subject a wager, and he takes it, and then he sees a coin flip or a die roll, that may alter his expectations the next time. If money changes hands, it may be 'rational' for him to regard the next win or lose in a new light. And so forth. We tried in various ways to eliminate such conditioning or learning. Still, might not just being asked questions, or answering them, lead to changes? Experiment showed that it did, systematically and surprisingly. I was not able, however, to produce a reasonable theory to explain the change. But then, what did the earlier experiments really show? They were designed to test a static theory when the assumption of stasis was evidently false.[17]

My conclusion from these experiments, and a hundred more of

[16] Donald Davidson and J. Marschak, 'Experimental Tests of a Stochastic Decision Theory'.
[17] For another account of this series of experiments, see Essay 12.

which I have read, is that they *can* be taken, if we want, as testing whether decision theory is true. But it is at least as plausible to take them as testing how good one or another criterion of preference is, on the assumption that decision theory is true.

Amos Tversky, a tireless and skilled worker in this area, seems to have come to much the same conclusion. He tells of some experiments done by him and Daniel Kahneman in Jerusalem.[18] They asked subjects to choose between A and B, and C and D, in the following situation:

Choice I: A = (1,000, 1/2, 0)   B = (400)
Choice II: C = (1,000, 1/10, 0)   D = (400, 1/5, 0)

Here A, for example, is a wager that is equally apt (subjectively) to pay $1,000 or nothing; B is simply receiving $400 (no risk). Tversky says almost all subjects picked B over A and C over D; the result held for naïve and educated subjects, and analogous results were obtained for different payoffs and probabilities. The trouble is, this pattern of preferences is incompatible with decision theory. Apparently.

But as Tversky points out, the claim that decision theory is wrong here depends on the assumption that the consequences under consideration are 'fully characterized by the relevant monetary values'. We need not accept this assumption. For example, he suggests that the reason subjects shy away from A is that the zero outcome should be described as missing out on the prospect of getting $400 for certain. Put differently: the lack of risk in B had a value of its own. Given *this* assumption, decision theory is not necessarily falsified by the results. As Tversky says,

The question of whether utility theory is compatible with the data or not . . . depends critically on the interpretation of the consequences . . . The key issue, therefore, is not the adequacy of the axioms, but rather the appropriateness of the interpretation of the consequences . . . In the absence of any constraints, the consequences can always be interpreted so as to satisfy the axioms.[19]

It may seem that I want to insist that decision theory, like the simple postulate that people tend to do what they believe will promote their ends, is necessarily true, or perhaps analytic, or that it states part of what we mean by saying someone prefers one alterna-

[18] Amos Tversky, 'A Critique of Expected Utility Theory: Descriptive and Normative Considerations'.   [19] Ibid., 171.

tive to another. But in fact I want to say none of these things, if only because I understand none of them. My point is sceptical, and relative. I am sceptical that we have a clear idea what would, or should, show that decision theory is false; and I think that compared to attributions of desires, preferences or beliefs, the axioms of decision theory lend little empirical force to explanations of action. In this respect, decision theory is like the theory of measurement for length or mass, or Tarski's theory of truth. The theory in each case is so powerful and simple, and so constitutive of concepts assumed by further satisfactory theory (physical or linguistic) that we must strain to fit our findings, or our interpretations, to preserve the theory. If length is not transitive, what does it mean to use a number to measure a length at all? We could find or invent an answer, but unless or until we do we must strive to interpret 'longer than' so that it comes out transitive. Similarly for 'preferred to'.

I think I have an argument to show that the main empirical thrust of an explanation of an action in decision theory, or of a reason explanation, does not come from the axioms of decision theory, or 'the assumption of rationality', but rather from the attributions of desires, preferences, or beliefs. Suppose, as I have been, that Hempel is right when he claims that every explanation states or implies an empirical generalization. Then an explanation tells a great many things that go beyond the case to be explained; the law plus other premises tells us that the event to be explained had to occur, or was likely to, but the law also tells us that *if* certain further things happen, there will be specified consequences. An explanation unconditionally predicts what it explains (in the sense that the sentence to be explained can be deduced from the law and the statement of antecedent conditions), and conditionally predicts endless further things. So by asking what an explanation of a particular event conditionally predicts we learn what sort of law is involved. Now ask yourself what conditional predictions flow from a reason explanation, let's say of Ford's compromise on the energy bill in order to curry favour with the voters. If the assumption of rationality is the important empirical law, then this explanation of Ford's action should tell us what (conditionally) to expect of all mankind. But if I am right, it mainly tells us what (conditionally) to expect of Gerald Ford. For what I have suggested is that the relevant generalizations are just the ones that express dispositions like wanting to curry favour with the voters, or believing that compromizing on the energy bill would make the voters like him. To me the case seems clear; knowing the explanation of Ford's action tells us a lot about him, but almost nothing about people generally.

If the assumption of rationality (or decision theory) had the empirical force that Hempel's account requires, the explanation of Ford's action would be like an explanation of why a particular small cube dissolved in warm coffee, an explanation that takes as premises the facts that if something is soluble it dissolves in liquids of a certain sort, that warm coffee is such a liquid, that all sugar is soluble, and that this cube is sugar. This explanation implies something about all sugar and all coffee ('all mankind'); but there is a shorter and less informative explanation available. This cube was soluble; soluble things dissolve in coffee, this cube was in coffee. This explanation tells us nothing about all sugar-kind, though it tells us a lot about this cube. I think reason explanations are of the second kind.

Explaining why something dissolved by reference to its solubility is not high science, but it isn't empty either, for solubility implies not only a generalization, but also the existence of a causal factor which accounts for the disposition: there is something about a soluble cube of sugar that causes it to dissolve under certain conditions. That this is so is not proved by the fact that it dissolves once. In the same way, no single action can prove that a disposition like a desire or belief exists; desires and beliefs, however short lived, cannot be momentary, which is why we typically learn so much from knowing about the beliefs and desires of an agent. This is the point, I suggest, where general knowledge of the nature of agents *is* important, general knowledge of how persistent various preferences and beliefs are apt to be, and what causes them to grow, alter and decay. Such knowledge is not used in giving reason explanations, but it is surely part of why reason explanations are so satisfying and informative.

Hempel set out to show that reason explanations do not differ in their general logical character from explanation in physics or elsewhere. My reflections reinforce this view. On one point I am not persuaded, however: the laws that are implicit in reason explanation seem to me to concern only individuals—they are the generalizations embedded in attributions of attitudes, beliefs, and traits. We do have nomic wisdom concerning mankind and womankind, but this wisdom does not, I have urged, directly enter into reason explanations. This wisdom mainly advises us about the nature of dispositions like desires and beliefs—particularly how long they are apt to last, and what causes them to alter (learning, perception). This knowledge does not make our reason explanations any stronger, but it does make them more valuable by letting us fit them into a larger scheme.

15 *Hume's Cognitive Theory of Pride*

Most of Book II of the *Treatise* is devoted to the study of what Hume calls the 'indirect passions'; these are emotions and attitudes that can be explained, Hume thinks, only by their causal relations with beliefs. Among the examples Hume gives are pride, humility, vanity, love, hatred, envy, pity, malice, esteem, benevolence, respect, and compassion. It is evident that there are serious inadequacies in Hume's theory, so it is understandable that most recent writers who have discussed Hume on the passions have focused on the flaws.[1] They have gone wrong, in my view, in rejecting the causal aspect of Hume's doctrine as if it were inseparable from the atomistic psychology. But what is far more lamentable is that Hume's genuine insights have been almost totally ignored. For if I am right these insights, when freed from Hume's epistemological machinery, can be restated and assembled into an intriguing and persuasive theory. In what follows I urge the merits of this theory and show how it can be extracted from the *Treatise*. I do not pretend that this is what Hume really meant; it is what he *should* have meant, and did inspire. Like Hume, I shall concentrate on pride.

Hume's account of pride is best suited to what may be called *propositional pride*—pride described by sentences like, 'She was proud that she had been elected president.' Hume more often speaks of being proud *of* something—a son, a house, an ability, an accomplishment—but it is clear from his analysis that cases of being proud of something (or taking pride in something, or being proud to

---

[1] Examples: P. L. Gardiner, 'Hume's Theory of the Passions', 31–42; Anthony Kenny, *Action, Emotion and Will*; Irving Thalberg, 'Emotion and Thought', 1–11; George Pitcher, 'Emotion', 326–46; William Alston, 'Moral Attitudes and Moral Judgments', 1–23.

do something) reduce to, or are based on, propositional pride. If Hume's theory is to cope with the other indirect passions, a propositional form must be found for each of them. (Thus Hume seems to explain loving *for a reason* rather than simply loving.)

It is commonly thought that in Hume's opinion, to be proud that one is clever (say) is just to experience a certain 'ultimate felt quality'. This would certainly be wrong, since it would provide no way of distinguishing between being proud that one is clever and being proud that one is kind to kangaroos. What Hume said is that pride is a 'simple and uniform impression'[2] which is 'an original existence, or, if you will, modification of existence, and contains not any representative quality, which renders it a copy of any other existence' (415); and this is consistent with holding that being proud *that* one is clever is a complex state of which the simple impression of 'pride' is only one element.

I do not defend Hume's use of the word 'pride', which does not seem to correspond to any use the word has in English; but this terminological aberration is irrelevant to the evaluation of Hume's theory of propositional pride, which concerns the conditions under which predicates like 'x is proud that that he is clever' apply. Hume is vulnerable, however, if he held that propositional pride always involves a particular feeling or sensation, for though there are sometimes characteristic frissons of pleasure that accompany prideful thoughts, such experiences are not necessary or typical. But Hume, perhaps, inconsistently, said that the impression of pride could become 'imperceptible' (276); that it might be so 'calm' as to 'cause no disorder in the soul', and even be confused with a 'determination of reason' (417); and that once a passion becomes a 'settled principle of action ... it commonly produces no longer any sensible agitation' (419). The legitimate role Hume's impression of pride plays in his finished theory is to account for a common affective element in propositional pride; since no atomistic perception could play this role, I shall attend to the role and let the atomism go.

Suppose a man is proud that he has a beautiful house; then his state of mind is caused, according to Hume, by a belief that he has a beautiful house. *What* the man takes pride in, that is, the fact that he

---

[2] David Hume, *A Treatise of Human Nature*, 277. Subsequent references to this work (in the Selby-Bigge edition) given the page number in parentheses.

has a beautiful house, is identical with the content of his belief: one could say that the belief *determines* the object of pride.[3]

Hume ignores the fact that an attribution of propositional pride is false unless the corresponding belief is true (except when he suggests that we can criticize a passion if it is 'founded' on a false belief (416)); and this seems reasonable, since it is unclear what interesting difference there is between pride founded on a true belief and an otherwise similar passion founded on a false belief.

There is a formal restriction on the belief that causes pride (and hence on *what* pride can be taken in): it must be a belief to the effect that the believer himself has some property. ('I have a beautiful house', 'My son is agile', 'I have good address in fencing' (279).) This is not quite how Hume explains the restriction. He says that the 'subject' of pride must be closely related to the proud person, and that some 'quality' must be thought to belong to the subject; in the example, the subject is the house, the quality beauty, and the relation between the subject and the proud person that of ownership. In special cases, the relation is that of identity, and the subject is the proud person himself (for example, if he is proud that he is clever). It is simpler to drop Hume's distinction, and insist that the belief always have the form 'I . . .' where the dots supply a predicate that may or may not contain a reference to some further object. This builds a belief in the appropriate relationship between self and the further object into the central causal belief, which is where it belongs. For a belief that a house is beautiful will not make a man proud unless he believes that it is his house (or is otherwise related to him). It is clear that the reference to self must be pronomial: Smith is not proud that Smith is president of the bank, but that *she* is president of the bank.

There is a further reason for revising Hume's formal condition. Hume claims that the property we attribute to the subject must be one toward which we have a positive attitude; so someone would not be proud that he had a beautiful house unless he thought beauty was good (or he had some other positive attitude toward beauty). But in fact this is not quite right. Someone could be proud that he

---

[3] This is not to say, of course, that the belief *is* the 'object of Pride'. All this talk is loose. I do not assume that 'the object of pride', 'what pride is taken in', 'the content of belief' refer to psychological entities of any kind. Of course the semantic analysis of intentional sentences, like those which attribute belief or propositional pride, may require objects such as propositions, sentences, or utterances.

had an ugly house, even while disliking ugliness, for he might consider it a virtue to be careless about his surroundings. The property the proud man must be positive about is not the property he believes the house has but the property he believes *he* has—in this case the property of being the owner of a beautiful (or ugly) house.

Hume nowhere, I think, calls the cause of pride a belief; he speaks, rather, of an idea which has two 'component parts' that must be in 'conjunction' to produce the passion; one part is the idea of the subject, the other is (the idea of) the 'quality' (279). Obviously this conjunction must amount to predication, for Hume speaks of the quality being 'placed upon' the subject (279, 285) or 'adhering' to it (289). His explanation of the sense in which passions may be unreasonable requires that they be 'founded' on 'judgements' or 'suppositions' (416). Beliefs for Hume are ideas that are 'vivid and intense'; 'the effect, then, of belief is to raise up a simple idea to an equality with our impressions, and bestow on it a like influence on the passions' (119). Finally, 'belief is almost absolutely requisite to the exciting our passions' (120).

The belief that one has a certain quality or property can produce pride only if one prizes possession of the property. This fact leads Hume to one of his most interesting claims: in order to be proud that one has a property, the property itself must quite independently cause an impression that 'resembles and corresponds' to pride (287). This resemblance does not make the separate impression a form of pride; the resemblance depends rather on the fact that both impressions are pleasant or positive. When Hume examines the properties which, when thought to be our own, produce pride, he finds that they 'concur in producing the sensation of pleasure' quite apart from their relation to pride: 'thus the beauty of our person, of itself, and by its very appearance, gives pleasure, as well as pride ... every cause of pride, by its peculiar qualities, produces a separate pleasure' (285).

It would be a mistake to suppose that in Hume's view it is only our own beauty or virtue that pleases us. On the contrary, the point of the 'separate' pleasure we take in the things that make us proud is that it *explains* our pride. That virtue and beauty produce pleasure generally is not for Hume a contingent matter: 'the very essence of virtue ... is to produce pleasure' (296), while pleasure is not only the 'necessary attendant' of beauty, but constitutes its 'very essence'

(299). 'Everything useful, beautiful or surprising, agrees in producing a separate pleasure' (301). This doctrine is not without its ambiguities, one of which is that Hume seems often to confuse what gives us pleasure and what we are pleased is the case. The first of these may explain the second, but it is the second that promises to help with the analysis of propositional pride. There is, however, a difficulty. Hume remarks that we are more apt to be proud of what is uniquely ours than of what we share with many others. But then we can hardly be pleased that others have the trait we are proud of. It would be more natural to say that a trait we are pleased that anyone should have is trait we should be pleased to have ourselves; but this trait would not normally be a source of pride.

What is needed to account for pride is the attitude of *approval*, or thinking well of, rather than being pleased. A man who is proud that he owns a beautiful house, or that he has designed one, may not be pleased that others own or design beautiful houses, but he is bound to approve of others, or think well of them, for such possessions or accomplishments. We can make this shift—from considerations of pleasure to the attitude of approval or thinking well—without doing violence to Hume's views. For while he held that 'To have the sense of virtue, is nothing but to *feel* a satisfaction of a particular kind from the contemplation of a character', he also held that

The very *feeling* constitutes our praise or admiration ... We do not infer a character to be virtuous, because it pleases: But in feeling that it pleases after such a particular manner, we in effect feel that it is virtuous. The case is the same as in our judgements concerning all kinds of beauty, and tastes, and sensations. Our approbation is imply'd in the immediate pleasure they convey to us. (471)

I shall interpret Hume, then, as maintaining that, if someone is proud that he exemplifies a certain property, then he approves of, or thinks well of, others for exemplifying the same property. This approval is not to be distinguished from holding that anyone who has the property is to that extent praiseworthy, estimable, or virtuous.[4]

The 'separate' passion which pride requires, and which I am interpreting as the attitude of approbation, plays a crucial role in Hume's account of how pride is produced. Here is a summary of

---

[4] Here, and elsewhere in this Essay, my view of Hume's theory is close to that of Páll Árdal in *Passion and Value in Hume's Treatise*. Unfortunately I had not seen Páll's excellent book when I wrote this Essay.

Hume's theory, more or less as he gives it: the cause of pride is a conjunction of the idea of a house, say, and a quality (beauty). The quality causes the separate and pleasant passion, which under the right conditions causes (by association) the similar pleasant passion of pride. The passion of pride itself always causes the idea of self to appear, and this idea must be related (causally, by association) to the idea of the object (the house) on which the quality is placed. In short, 'That cause, which excites the passion [pride], is related to the object [self], which nature has attributed to the passion; the sensation, which the cause separately produces, is related to the sensation of the passion: From this double relation of ideas and impressions, the passion is derived' (286).

Hume wants to make this welter of causal relations sound simple: he says, 'The pleasure, therefore, with the relation to self must be the cause of the passion', or 'the pleasure produces the pride by a transition along related ideas', (301) or, 'there is a transition of affections along the relation of ideas' (306). I think Hume was right: there is a fairly simple and intelligible order beneath what he saw as the play of ideas and impressions, and he provided us with most of what we need to chart the basic structure.

To begin with a difficulty: Hume wants (correctly, I think) to distinguish between the belief that causes pride and the 'object' to which pride is directed (278). The latter is always the idea of self, while the total cause is complex, but Hume does not depend on this alone to distinguish cause and object. He also wants to keep separate the first component of the cause, even when this is the idea of self (as it must be when someone is proud of his beauty or wit), and the object. The distinction rests, therefore, not on what the idea represents, but on its causal relation to pride: pride '... is a passion placed betwixt two ideas, of which one produces it, and the other is produced by it' (278). The problem is to understand the final claim, that pride causes the idea of self.

Hume is defensive when he tries to explain the relation between the pleasurable sensation of pride and its 'object', the self. He says that pride and humility are 'determined to have self for their *object*' by an 'original' property of the mind; this is 'the distinguishing characteristic of these passions' (280). 'Tis absolutely impossible, from the primary constitution of the mind, that these passions should ever look beyond self ... For this I pretend not to give any reasons; but consider such a peculiar direction of the thought as an original quality' (286). Pride produces the idea of self just as the nose and palate are disposed to convey the sensations that they do to the mind; 'All this needs no proof' (287).

The problem for Hume is this: if the simple impression of pride always (or perhaps necessarily) produces the idea of self, then what role is there left for the relation of ideas of which he makes so much, the one idea being part of the cause (the 'subject') and the other the idea of self that accompanies pride? On Hume's psychological theory, the association of these ideas might increase the vivacity of the subsequent one, but there is no way the association could strengthen the ties, causal or otherwise, between pride and its object. Hume wants the 'transition along related ideas' to be part of the explanation of how generalized pleasure or approbation is 'transfused' into pride (301, 338), for, if it is not, there is no reason why the belief that causes pride should concern oneself; but his causal machinery does not assign an intelligible role to the relation of ideas.

Hume's official solution is necessarily inadequate because his psychological apparatus cannot yield a serious account of judgement or belief. But if we allow ourselves the ordinary concepts of belief and judgement (as Hume often lapses into doing), we can put Hume's central insight in acceptable form. Notice, first, that according to Hume the impression of pride not only 'resembles and corresponds' to the generalized impression of approval towards the property which, when believed to be our own, makes us proud (287); it is somehow 'inseparable' from it (289). Indeed, 'in the transition the one impression is so much confounded with the other, that they become in a manner undistinguishable' (331). The former impression we made good sense of when we remembered that for Hume such an impression may be the feeling involved when we *feel* a satisfaction of a particular kind from the contemplation of a character. The very *feeling* constitutes our praise.' If something pleases us, we 'feel that it is virtuous'. The impression in this case is the affective aspect of 'approbation' or a 'judgement concerning . . . beauty, and tastes, and sensations' (471). We may, then, interpret Hume's 'simple impression' of pride as the affective aspect of self-approbation, or a judgement that oneself is virtuous, praiseworthy, beautiful, or otherwise endowed in a positive way. And Hume often does describe what it is like to be proud that one has some trait in just such terms: he contrasts the pleasure of a guest at a lavish feast with the 'self-applause' of the host (290); he points out that, 'We fancy ourselves ... more virtuous or beautiful, when we appear so to others' (292), and remarks that there is nothing immoral in pride when we 'receive a pleasure from reflecting on a generous action' (298). He does not distinguish the pride of the swan or peacock from 'the high idea he has entertained of himself' (326). When Hume defines pride as 'that agreeable impression, which arises in the mind, when the view either of our virtue, beauty, riches or power makes us satisfied with ourselves' (297) he does not intend to separate the impression from the satisfaction.

Self-approval, self-esteem, self-applause, constitute pride. We may follow Hume in abstracting from self-approval two elements, approval, and the reference to self, without accepting the doctrine that these elements are atoms of experience, and without having to identify one of the atoms with the 'pleasant impression' of pride. Shorn of these assumptions, the basic structure of pride and its etiology as Hume saw them is clear: the cause consists, first, of a belief concerning oneself, that one has a certain trait, and, second, of an attitude of approbation or esteem for anyone who has the trait. Together these result in self-approval or self-esteem—what is normally called pride.

It is now evident why Hume insisted on distinguishing the idea of self as part of the cause and the idea of self toward which pride is directed, for the belief that causes pride is separate from what it causes. One can also see why Hume wanted to distinguish the impression of pride from the impression of general approval, for though both are forms of approval, one is an aspect of a general attitude and the other an aspect of the self-approbation the general attitude helps cause.

The causal relation echoes a logical relation. Hume equates approbation with a judgement of merit. The self-approbation that is pride may then be expressed by a judgement that one is praiseworthy (I do not insist on this particular word). The causes of pride are a judgement that everyone who exemplifies a certain property is praiseworthy and a belief that one exemplifies that property oneself. The causes of pride are thus judgements that logically imply the judgement that is identical with pride. Hume's 'double relation' of ideas and impressions is his way of explaining the causal and logical relations between pride and the attitudes and beliefs on which it is based. There are two references to self, one in a premise and one in the conclusion; two attributions of merit, one in a premise and one in the conclusion.

The theory just outlined needs a major revision. The logic of pride, as stated, is a syllogism in Barbara: All who own a beautiful house are praiseworthy, I own a beautiful house, therefore I am praiseworthy. The trouble is that the middle term—'owns a beautiful house'—drops out in the conclusion, although it is needed there for an account of propositional pride. I am proud that I own a beautiful house; the associated judgement should be something like, I am praiseworthy for being the owner of a beautiful house, or I am to be esteemed in that I own a beautiful house. Such a conclusion is really all that makes sense; however much we esteem somebody for owning a beautiful house, we do not for that reason alone hold him estimable (he may have bought the house with embezzled funds). Of course, owning a beautiful house might cause someone to approve of himself generally—to become a proud man; but this case is not the one we set out to explain, and this man's pride would outreach his reason for being proud.

Since the belief and attitudes that lead to pride, as I have formulated them, do justify unconditional pride, the formulation must be wrong. The error is in the analysis of the attitude: I do not simply approve, or hold meritorious, all who own beautiful houses; rather, I approve of them, or hold them meritorious, *in that* they own beautiful houses. My approval or judgement is prima facie in character and is given only relative to its ground. From such approval, and the belief that I own a beautiful house, all I can conclude is that I am meritorous in that I own a beautiful house. This is the form of judgement that expresses being proud that I own a beautiful house. Hume appreciates the fact that the idea of the house must appear not only in the cause of pride, but also among the objects to which pride is directed, when he says, 'pride has in a manner two objects, to which it directs our view' (292); in the case at hand these are the house and oneself.

The theory of propositional pride that I have extracted from the *Treatise* shows that someone who is proud always has his reasons; the cause of his pride rationalizes it. As a result, giving the belief and attitudes on which pride is based explains the pride in two ways: it provides a causal explanation, and it gives the person's reasons for being proud. A full description of what it is that a person is proud of tells all that is needed to reconstruct his reasons, but a less than full description or an unexpected attitude can call for further specification. ('Why are you so proud of having got *down* Mt. Everest?' 'Well, I did it on skis.' 'Why are you proud that you never laugh?' 'Laughter is an infirmity of human nature.')

Hume will not allow that pride or any other passion is based on reason alone; but this is not to deny that some passions (the 'indirect', or propositional passions) are based on *reasons*. Hume's point is rather that if a passion is based on reasons (he does not use the word this way) at least one of the reasons must itself be, or be based on, a passion (416).

This familiar Humean doctrine, that 'reason is, and ought only to be the slave of the passions', contrasts, in an interesting matter of emphasis, with what Hume says about the cause of pride. In the general discussion of the roles of reason and passion in determining action, or deciding what is moral, Hume emphasizes the passive role of reason in contrast to the active role of passion. 'An active principle,' he says, 'can never be founded on an inactive; and if reason be inactive in itself, it must remain so in all its shapes and appearances' (457). 'Reason and judgement may, indeed, be the mediate cause of an action, by prompting, or by directing a passion' (462); in the same way, in the 'double relation' that causes pride, reason is requisite 'only to make one passion produce another' (420). Why then does Hume say *the* cause of pride is the belief, not the 'active' passion that is also required? The answer is, I think, simple. Both belief and attitude, reason and passion, are necessary to cause pride. But the relevant attitudes—the approval of beauty, strength, wit, power, and possessions—are, Hume thinks, universal (281). Where men differ is in their gifts, and hence in what they believe to be their gifts; this is therefore what needs to be mentioned in explaining pride, even if it is the 'inactive' principle.

The fact that someone who is proud that he has some characteristic has reasons does not show that such pride is always reasonable. The pride may, for example, be out of proportion to its grounds (416, 419). The grounds may be open to criticism: the relevant belief may be false or unreasonable (416), or the general attitude may be based in turn on unwarranted beliefs, or an injudicious weighting of considerations. So a man might be proud of his military victories, even while finding more to disapprove of than to admire in them. Since we all like to think well of ourselves, there is a temptation to reduce dissonance by persuading ourselves we have virtues we lack (321) or by deciding that the traits we have are virtues and not vices (292). Pride is open to criticism, then, but the criticism must concern the whole constellation of belief and attitude that is its direct source, since whatever forces for irrationality there are must operate to create the whole constellation (304, 396). Hume rightly, I think, considers this to be the strongest argument in favour of his theory (301–2, 335).

Now I want, much too briefly, to consider how this theory stands up to criticism. According to Alston, we must reject the Humean view that an emotion is a 'qualification of consciousness by some ultimate felt quality' and accept instead the view that being in an emotional state entails making, or being disposed to make, a kind of judgement.[5] The theory I have constructed identifies the state someone is in if he is proud that *p* with his having the attitude of approving of himself because of *p*, and this in turn (following Hume) I have not distinguished from judging or holding that one is praiseworthy because of *p*. Here again is what Hume says: 'by *pride* I understand the agreeable impression, which arises in the mind, when the view either of our virtue, beauty, riches or power makes us satisfied with ourselves' (297). Elsewhere, he identifies pride with 'a certain satisfaction in ourselves, on account of some accomplishment or possession, which we enjoy'.[6] In fact, Hume can be quoted on both sides: he does say that pride is a 'simple impression', but he also defines this impression in terms of its relations to beliefs and the object to which it is 'directed'. It seems to me a relatively trivial complaint against Hume that he identified pride with only one element in the structure in which he embedded it. The interesting question is whether the structure as a whole suffices to give an account of propositional pride. And here it is appropriate to agree with the critics to this extent: Hume surely did often, and characteristically, assert that a pleasant feeling, or a feeling of pleasure of a certain sort, was essential to pride, whereas no such feeling is essential; and, more important, such an element does not help in analysing an attitude of approval, or a judgement. On the other hand, since Hume thought a feeling of this kind was required for the analysis of approval, or of evaluative judgement, there is justifica-

---

[5] 'Moral Attitudes and Moral Judgments', 13, 14.
[6] David Hume, 'A Dissertation on the Passions', sect. ii.

tion for allowing approval or judgement to feature in Hume's theory, especially since Hume admitted that the passions may be so 'calm' or habitual as to be unnoticeable, or even 'very readily taken for the determinations of reason'.

Analogous remarks apply to Kenny's claim that 'Hume denies very explicitly the intensionality of the passions';[7] he quotes this passage: 'A passion is an original existence ... and contains not any representative quality, which renders it a copy of any other existence or modification. When I am angry, I am actually possessed with the passion, and in that emotion have no more a reference to any other object, than when I am thirsty, or sick, or more than five feet high' (415). The criticism is confused. What Hume *called* the passion had no 'representative quality'; but the pattern of elements he called on to make a passion what it is certainly did. The valid criticism is that what Hume called the passion has no place in the pattern.

Of course it also may be that the pattern is wrong. Hume was wrong to suppose that the state of being proud causes the idea of self to which it is directed; that idea is a constituent of the state. But suppose we allow ourselves to talk seriously of the intensional object of pride—the proposition as a whole. Was Hume also wrong about *its* relation to pride? It seems to me we cannot say this. Someone might argue that it is the belief that causes pride which supplies the propositional object in Hume's theory, and Hume mistakenly thinks this belief is only contingently related to the state of pride. (Kenny seems to use this argument.)[8] But this criticism cannot touch Hume, since Hume distinguishes between the cause of pride and what it is directed to. I think, as I said above, that we do best not to talk of the object of pride as if it were a psychological entity of some kind (semantics may require an object to correspond to the sentence ruled by 'is proud that', or perhaps the that-clause); but if we do want to talk of the object of pride, then as good a phrase as any for the relation between pride and its object is Hume's: 'directed to'.

Hume is often berated for having made the relation between pride and the relevant belief causal; and philosophers who have not mentioned Hume have agreed with the point.[9] The argument can-

[7] *Action, Emotion and Will*, 25.
[8] Ibid., 71.
[9] For example, B. A. O. Williams, 'Pleasure and Belief', 57–72.

not, as we have just seen, be made to depend on the fact that the relation between pride and its object is not causal, since it is possible to distinguish between the cause and the object. The more standard argument is that the relation between belief and pride cannot be causal because causal relations are contingent while the relation between belief and pride is necessary. At least, it is said, this may be argued against Hume, for *he* surely held that causal relations are contingent. The historical question is too large for us now; I do not myself think it can be cleared up definitively, since Hume's 'ideas' sometimes correspond to sentences and sometimes to referring phrases. But whatever we decide about Hume, there is not a good argument to show that causal relations rule out necessary connections. According to my dictionary, snowblink is a white luminosity on the underside of clouds caused by the reflection of light from a snow surface. This is a necessary truth, though a truth about a causal relation. In the same way, I think it is necessary that someone's being proud that *p* be caused by (among other things) a belief that *p*.[10] Some of Hume's remarks suggest that he thought this too; other of his remarks don't.

There is a further argument that is usually used by those who oppose, or question, the view that belief is a causal factor in propositional pride, and it is that, if the relation is causal, then certain present-tense self-attributions of pride could not have the kind of authority, or relative immunity to refutation by counter-evidence, that they do. Thus, 'I am proud because for once I managed to keep my mouth shut', would seem to be a causal claim; but then why do we (correctly) go on the presumption that such claims, when honest, are right?

How is this argument to be stated? If someone knows he is proud that he has a beautiful house, then he knows that his pride was caused by a belief that he has a beautiful house. This is not open to empirical question because, according to the theory, it is a necessary truth. But, it is said, if the relation is causal, then cause and effect can be redescribed in a way that makes the causal claim contingent. This is true; but if cause and effect are redescribed, let's say in physiological terms, we would admit that there is no first-person authority.

[10] The case for this view is well argued in J. R. S. Wilson, *Emotion and Object*, and Robert M. Gordon, 'The Aboutness of Emotion'. Both authors rely in part on my treatment of actions and reasons, and of causality, but neither presents the theory of emotion suggested here.

The only intelligible way to state the argument, I think, is to ask how someone can speak with special authority when he attributes propositional pride to himself. For when he says, 'I am proud that I have a beautiful house' he is making a causal claim, and what he implies caused his state of pride, namely a belief, is certainly distinct from what it is supposed to have caused. Just as the right question to ask is not, 'How do you know that the snowblink we now see is caused by the reflection of light from snow?' but rather, 'How do you know that what we now see is snowblink?' This last is a question whose answer calls for general causal knowledge, an answer that can be questioned or supported by pointing to parallel negative or positive instances. Why can't the same sort of doubt arise with pride?

The same *sort* of doubt can arise. A person can easily be wrong about exactly what it is that he is proud of; whether it is the cost or the cut of his clothes, say, or his skill at tennis rather than his score. If we suppose he knows he is proud of something, and he knows (or has special authority with respect to) what he believes, the question which of several related beliefs was the cause can not only be troubling, but it also can be questioned by reference to parallel cases. Still, the fact that claims to know what one is proud of can be wrong, and can be questioned, does not show that we do not have special authority with respect to them. The source of this authority springs from the nature of human interpretation of human thoughts, speech, intentions, motives, and actions. If we are to understand any of these, make sense of them *as* thoughts and speech and actions, we must find a way to read into them a pattern subject to complex constraints. Some of these constraints are logical, and some are causal; some are both. What operates as a constraint on the interpreter amounts to a bestowal of authority on the person interpreted; when he honestly expresses his motives, beliefs, intentions, and desires an interpreter must, if he wants to understand, interpret the speech in a way that makes sense not only of what the speaker says but also of what he believes and wants and does. Making such sense requires that causality and rationality go hand in hand in important cases: perception, intentional action, and the indirect passions are examples. People are in general right about the mental causes of their emotions, intentions, and actions because as interpreters we interpret them so as to make them so. We must, if we are to interpret at all.

Bibliographical References

Alston, William, 'Moral Attitudes and Moral Judgments', *Noûs*, 2 (1968), 1–23.

Anscombe, G. E. M., *Intention*. Blackwell, Oxford. (1959).

Anscombe, G. E. M., 'Thought and Action in Aristotle', in *New Essays on Plato and Aristotle*, ed. R. Bamborough. Routledge and Kegan Paul, London (1965), 143–58.

Aquinas, *Summa Theologica*.

Árdal, P., *Passion and Value in Hume's Treatise*. Edinburgh University Press, Edinburgh (1966).

Armstrong, D., 'Acting and Trying', *Philosophical Papers*, 2 (1973), 1–15.

Armstrong, D., 'Beliefs and Desires as Causes of Action: A Reply to Donald Davidson', *Philosophical Papers*, 4 (1975), 1–8.

Aristotle, *Nichomachean Ethics*.

Atwell, J. E., 'The Accordion-Effect Thesis', *Philosophical Quarterly*, 19 (1969), 337–42.

Augustine, *Confessions*.

Austin, J. L., 'Unfair to Facts', in *Philosophical Papers*. Clarendon Press, Oxford (1961), 102–22.

Austin, J. L., 'A Plea for Excuses', in *Philosophical Papers*. Clarendon Press, Oxford (1961), 123–52.

Austin, J. L., 'Ifs and Cans', in *Philosophical Papers*. Clarendon Press, Oxford (1961), 153–80.

Austin, J. L., 'Three Ways of Spilling Ink', *Philosophical Review*, 75 (1966), 427–40.

Ayers, M. R., *The Refutation of Determinism*. Methuen, London. (1968).

Baier, K., *The Moral Point of View*. Cornell University Press, Ithaca (1958)

Bennett, D., 'Action, Reason and Purpose' *Journal of Philosophy*, 62 (1965), 85–95.

Berofsky, B., 'Causality and General Laws', *Journal of Philosophy*, 63 (1966), 148–57.

Brand, M., 'Danto on Basic Actions', *Noûs*, 2 (1968), 187–90.

Brandt, R., and Kim, J., 'The Logic of the Identity Theory', *Journal of Philosophy*, 64 (1967), 515–37.

Bullard, E. C., 'The Detection of Underground Explosions', *Scientific American*, 215 (1966), 19–29.

Burks, A., 'The Logic of Causal Propositions', *Mind*, 60 (1951), 363–82.

Butler, S., *Sermons*.

Cargile, J., 'Davidson's Notion of Logical Form', *Inquiry*, 13 (1970), 129–39.

Castañeda, H.-N., 'Comments', in *The Logic of Decision and Action,* ed. N. Rescher. University of Pittsburgh Press, Pittsburgh (1967), 104–12.

Chisholm, R., *Perceiving*, Cornell University Press, Ithaca (1957).

Chisholm, R., 'J. L. Austin's *Philosophical Papers*', *Mind*, 73 (1964), 1–26.

Chisholm, R., 'The Descriptive Element in the Concept of Action', *Journal of Philosophy*, 61 (1964), 613–24.

Chisholm, R., 'The Ethics of Requirement', *American Philosophical Quarterly*, 1 (1964), 1–7.

Chisholm, R., 'Freedom and Action', in *Freedom and Determinism,* ed. K. Lehrer. Random House, New York (1966), 28–44.

Chisholm, R., 'Comments', in *The Logic of Decision and Action*, ed. N. Rescher. University of Pittsburgh Press, Pittsburgh (1967), 113–14.

Chisholm, R., 'Events and Propositions', *Noûs*, 4 (1970), 15–24.

Chisholm, R., 'States of Affairs Again', *Noûs*, 5 (1971), 179–189.

Church, A., *Introduction to Mathematical Logic*. Princeton University Press, Princeton, New Jersey (1956).

Churchland, P., 'The Logical Character of Action Explanations', *Philosophical Review*, 79 (1970), 214–36.

Danto, A., 'What We Can Do', *Journal of Philosophy*, 60 (1963), 435–45.

Danto, A., 'Basic Actions', *American Philosophical Quarterly*, 2 (1965), 141–8.

Danto, A., 'Freedom and Forbearance', in *Freedom and Determinism*, ed. K. Lehrer. Random House, New York (1966), 45–64.

Davidson, D., Suppes, P., Siegel, S., *Decision-Making: An Experimental Approach*. Stanford University Press, Stanford (1957). [Reprinted by University of Chicago Press, Chicago (1977).]

Davidson, D., 'Theories of Meaning and Learnable Languages', in *Proceedings of the 1964 International Congress of Logic, Methodology and the Philosophy of Science*. North-Holland Publishing Company, Amsterdam, (1965), 383–94.

Davidson, D., 'Truth and Meaning', *Synthèse* 17 (1967), 304–23.

Davidson, D., 'On Saying That', *Synthèse* 19 (1968–9), 130–46.

Davidson, D., 'True to the Facts', *Journal of Philosophy*, 66 (1969), 748–64.

Davidson, D., 'On the Very Idea of a Conceptual Scheme', *Proceedings and Addresses of the American Philosophical Association*, 67 (1973–4), 5–20.

Dray, W., *Laws and Explanation in History*. Oxford University Press, London (1957).

Dretske, F. I., 'Can Events Move?', *Mind*, 76 (1967), 479–92.

Ducasse, C. J., *Causation and the Types of Necessity*. University of Washington Press, Seattle (1924).

Ducasse, C. J., 'Explanation, Mechanism and Teleology', *Journal of Philosophy*, 22 (1925), 150–5.

Ducasse, C. J., *Nature, Mind, and Death*. Open Court, La Salle, Illinois (1951).

Ducasse, C. J., 'Critique of Hume's Conception of Causality', *Journal of Philosophy*, 63 (1966), 141–8.

Feigl, H., 'The "Mental" and the "Physical"', in *Minnesota Studies in the Philosophy of Science*, 2, ed. H. Feigl, M. Scriven, and G. Maxwell. University of Minnesota Press, Minneapolis (1958), 370–497.

Feinberg, J., 'Action and Responsibility', in *Philosophy in America*, ed. M. Black. Cornell University Press, Ithaca (1965), 134–160.

Feinberg, J., 'Causing Voluntary Actions', in *Metaphysics and Explanation*, ed. W. H. Capitan and D. D. Merrill. University of Pittsburgh Press, Pittsburgh (1965), 29–47.

Føllesdal, D., 'Quantification into Causal Contexts', in *Boston*

Studies in the Philosophy of Science, 2, ed. R. S. Cohen and M. Wartofsky. M.I.T. Press, Cambridge, Mass. (1960), 263–274.

Føllesdal, D., 'Quine on Modality', in *Words and Objections: Essays on the Work of W. V. Quine*, ed. D. Davidson and J. Hintikka. D. Reidel Publishing Company (1969), 175–85.

Frankfurt, H., 'Alternate Possibilities and Moral Responsibility', *Journal of Philosophy*, 66 (1969), 829–39.

Gardiner, P. L., 'Hume's Theory of the Passions', in *David Hume: A Symposium*, ed. D. F. Pears. Macmillan, London (1963), 31–42.

Geach, P. T., *Mental Acts*. Routledge and Kegan Paul, London (1957).

Goldman, A., *A Theory of Human Action*. Prentice-Hall, New York (1970).

Goodman, N., 'Comments', *Journal of Philosophy*, 63 (1966), 328–31.

Goodman, N., 'Two Replies', *Journal of Philosophy*, 64 (1967), 286.

Gordon, R. M., The Aboutness of Emotion', *American Philosophical Quarterly*, 11 (1974), 27–36.

Gorovitz, S., 'Causal Judgments and Causal Explanations', *Journal of Philosophy*, 62 (1965), 695–710.

Grice, H. P., 'Intention and Uncertainty', *British Academy Lecture*, Oxford University Press, London (1971).

Hampshire, S., *Thought and Action*. Chatto and Windus, London (1959).

Hampshire, S., *Freedom of the Individual*. Princeton University Press, Princeton, New Jersey (1965).

Hare, R. M., *The Language of Morals*. Clarendon Press, Oxford (1952).

Hare, R. M., *Freedom and Reason*. Clarendon Press, Oxford (1963).

Hart, H. L. A., and Honoré, A M., *Causation in the Law*. Clarendon Press, Oxford (1959).

Hedman, C., 'On the Individuation of Action', *Inquiry*, 13 (1970), 125–8.

Hempel, C. G., *Aspects of Scientific Explanation*. Free Press, New York (1965).

Hempel, C. G., 'Rational Action', in *Proceedings and Addresses of the American Philosophical Association.* The Antioch Press, Yellow Springs, Ohio (1962), 5–24.

Honoré, A. M. and Hart, H. L. A., *Causation in the Law.* Clarendon Press, Oxford (1959).

Hume, D., *A Treatise of Human Nature,* ed. L. A. Selby-Bigge. Clarendon Press, Oxford (1951).

Hume, D., 'A Dissertation on the Passions', in *Four Dissertations.* A. Millar, London (1757) [reproduced in facsimile by Garland Publishers, New York (1970)], 121–81.

Jeffrey, R. C., 'Goodman's Query', *Journal of Philosophy,* 63 (1966), 281–8.

Kant, I., *Fundamental Principles of the Metaphysics of Morals* trans. T. K. Abbott. Longman, Green and Co. (1909).

Kenny, A. J. P., *Action, Emotion and Will.* Routledge and Kegan Paul, London (1963).

Kim, J., 'On the Psycho-Physical Identity Theory', *American Philosophical Quarterly,* 3 (1966), 277–85.

Kim, J., and Brandt, R., 'The Logic of the Identity Theory', *Journal of Philosophy,* 64 (1967), 515–37.

Landesman, C., 'Actions as Universals', *American Philosophical Quarterly,* 6 (1969), 247–52.

Lehrer, K., 'Ifs, Cans and Causes', *Analysis,* 20 (1960), 122–4.

Lehrer, K., 'An Empirical Disproof of Determinism', in *Freedom and Determinism,* ed. K. Lehrer. Random House, New York (1966), 175–202.

Lemmon, E. J., 'Moral Dilemmas', *Philosophical Review,* 71 (1962), 139–58.

Lemmon, E. J., 'Comments', in *The Logic of Decision and Action,* ed. N. Rescher. University of Pittsburgh Press, Pittsburgh (1967), 96–103.

Levin, H., *The Question of Hamlet.* Oxford University Press, New York (1959).

Lewis, D. K., 'An Argument for the Identity Theory', *Journal of Philosophy,* 63 (1966), 17–25.

Luce, D. R., 'Mind–Body Identity and Psycho-Physical Correlation', *Philosophical Studies,* 17 (1966), 1–7.

Mackie, J. L., 'Causes and Conditions', *American Philosophical Quarterly,* 2 (1965), 245–64.

Malcolm, N., 'Scientific Materialism and the Identity Theory', *Dialogue,* 3 (1964–5), 115–25.

Martin, R., 'On Events and Event-Descriptions', in *Fact and Existence*, ed. J. Margolis. Basil Blackwell, Oxford (1969), 63–73.

Martin, R., 'Reply', in *Fact and Existence*, ed. J. Margolis. Basil Blackwell, Oxford (1969), 97–109.

Matthews, G., 'Mental Copies', *Philosophical Review*, 78 (1969), 53–73.

Melden, A. I., *Free Action*. Routledge and Kegan Paul (1961).

Mill, J. S., *Utilitarianism*.

Mill, J. S. *A System of Logic*.

Montague, R., 'Pragmatics', in *Contemporary Philosophy*, ed. R. Klibansky. La Nuova Italia Editrice, Florence, (1968), 102–22.

Montague, R., 'On the Nature of Certain Philosophical Entities', *Monist*, 53 (1968), 159–93.

Moravcsik, J., 'Strawson and Ontological Priority', in *Analytical Philosophy*, Second Series, ed. R. J. Butler. Barnes and Noble, New York (1965), 106–19.

Morgenstern, O. and von Neumann, J., *Theory of Games and Economic Behaviour*. Second Edition, Princeton University Press, Princeton, New Jersey (1947).

Mosteller, F., and Nogee, P., 'An Experimental Measurement of Utility', *Journal of Political Economy*, 59 (1951), 371–404.

Nagel, T., 'Physicalism' *Philosophical Review*, 74 (1965), 339–56.

Nogee, P., and Mosteller, F., 'An Experimental Measurement of Utility', *Journal of Political Economy*, 59 (1951), 371–404.

Nowell-Smith, P. H., *Ethics*. Penguin Books, London (1954).

Pap, A., 'Disposition Concepts and Extensional Logic', in *Minnesota Studies in the Philosophy of Science*, 2, ed. H. Feigl, M. Scriven, and G. Maxwell. University of Minnesota Press, Minneapolis, (1958), 196–224.

Pears, D. F., 'Ifs and Cans', *Canadian Journal of Philosophy*, 1 (1972), 249–74.

Peters, R. S., *The Concept of Motivation*. Routledge and Kegan Paul, London (1958).

Pitcher, George, 'Emotion', *Mind*, 64 (1965), 326–46.

Plato, *Philebus*.

Plato, *Protagoras*.

Quine, W. V. O., *Word and Object*. M. I. T. Press, Cambridge, Mass. (1960).

Quine, W. V. O., 'Ontological Reduction and the World of Numbers', in *Ways of Paradox*. Random House, New York (1966), 199–207.

Quine, W. V. O., 'Existence and Quantification', in *Fact and Existence*, ed. J. Margolis. Basil Blackwell, Oxford (1969), 1–17.

Ramsey, F. P., 'Facts and Propositions', reprinted in *Foundations of Mathematics*, Humanities Press, New York (1950), 138–55.

Ramsey, F. P., 'Truth and Probability', reprinted in *Foundations of Mathematics*. Humanities Press, New York (1950), 156–98.

Reichenbach, H., *Elements of Symbolic Logic*. Macmillan Co., New York (1947).

Rescher, N., 'Aspects of Action', in *The Logic of Decision and Action,* ed. N. Rescher. University of Pittsburgh Press, Pittsburgh (1967), 215–20.

Russell, B., *My Philosophical Development*. George Allen and Unwin, London (1959).

Ryle, G., *The Concept of Mind*. Barnes and Noble, New York (1949).

Santas, G., 'The Socratic Paradoxes', *Philosophical Review*, 73 (1964), 147–64.

Santas, G., 'Plato's *Protagoras* and Explanations of Weakness', *Philosophical Review*, 75 (1966), 3–33.

Scheffler, I., *The Anatomy of Inquiry*. Knopf, New York (1963).

Sellars, W., 'Metaphysics and the Concept of a Person', in *The Logical Way of Doing Things*, ed. K. Lambert. Yale University Press, New Haven (1969), 219–54.

Shoemaker, S., 'Ziff's Other Minds', *Journal of Philosophy*, 62 (1965), 587–9.

Shorter, J. M., 'Causality, and a Method of Analysis', in *Analytical Philosophy*, Second Series, ed. R. J. Butler. Barnes and Noble, New York (1965), 145–57.

Siegel, S., Davidson, D., Suppes, P., *Decision-Making: An Experimental Approach*. Stanford University Press, Stanford (1957).

[Reprinted by University of Chicago Press, Chicago (1977).]

Singer, M., 'Moral Rules and Principles', in *Essays in Moral Philosophy,* ed. A. I. Melden. University of Washington Press, Seattle (1958), 160–97.

Smart, J. J. C., 'Sensations and Brain Processes', *Philosophical Review*, 68 (1959), 141–56.

Strawson, P. F., *Individuals*. Methuen, London (1959).

Strawson, P. F., 'Contribution to a Symposium on "Determinism"', in *Freedom and the Will* ed. D. F. Pears. St. Martin's Press, London (1963), 48–68.

Stoutland, F., 'Basic Actions and Causality', *Journal of Philosophy*, 65 (1968), 467–75.

Suppes, P., Davidson, D., Siegel, S., *Decision-Making: An Experimental Approach*. Stanford University Press, Stanford (1957). [Reprinted by University of Chicago Press, Chicago (1977).]

Taylor, C., 'Mind–Body Identity, A Side Issue?', *Philosophical Review*, 76 (1967), 201–13.

Taylor, R., *Action and Purpose*. Prentice-Hall, Englewood Cliffs, New Jersey (1966).

Thalberg, I., 'Emotion and Thought', *American Philosophical Quarterly*, 1 (1964), 1–11.

Thalberg, I., 'Do We Cause our Own Actions?', *Analysis*, 27 (1967), 196–201.

Thalberg, I., 'Verbs, Deeds and What Happens to Us', *Theoria*, 33 (1967), 259–77.

Tversky, A., 'A Critique of Expected Utility Theory: Descriptive and Normative Considerations', *Erkenntnis*, 9 (1975), 163–74.

Vendler, Z., 'Effects, Results and Consequences', in *Analytic Philosophy* ed. R. J. Butler. Barnes and Noble, New York (1962), 1–14.

Vendler, Z., 'Comments', *Journal of Philosophy*, 62 (1965), 602–4.

Vendler, Z., 'Causal Relations', *Journal of Philosophy*, 64 (1967), 704–13.

Vickers, J. M., 'Characteristics of Projectible Predicates', *Journal of Philosophy*, 64 (1967), 280–5.

von Neumann, J. and Morgenstern, O., *Theory of Games and Economic Behavior*. Second Edition, Princeton, New Jersey (1974).

von Wright, G. H., *Norm and Action*. Routledge and Kegan Paul, London (1963).

Wallace, J. R., 'Goodman, Logic, Induction', *Journal of Philosophy*, 62 (1966), 310–28.

Williams, B. O., 'Pleasure and Belief', *Proceedings of the Aristotelian Society*, supplementary volume 33 (1959), 57–72.

Williams, C., *The Figure of Beatrice*. Faber and Faber, London (1943).

Wilson, J. R. S., *Emotion and Object*. Cambridge University Press, New York (1972).

Winch, P., *The Idea of a Social Science*. Routledge and Kegan Paul, London (1958).

Wittgenstein, L. W., *Blue and Brown Books*. Basil Blackwell, Oxford (1958).

Index

accordion effect, 53–61
actions:
  described in terms of causes, 10, 76
  descriptions of, 4, 8, 56–61, 71, 109–10, 193–6
  primitive, 49–61
  rationalizations of, 3–4
  sentences about as quasi-intensional, 5, 122, 127, 147–8, 193–6
  see also accordion effect; agency; causal explanation of action; intention; logical form; primary reasons
adverbial modification, 105–8, 114, 118–19, 155–60, 167, 185–6, 202
agency, 43–62, 114, 129
  accordion effect, 53–61
  agent causality, 19, 52–5, 60
  causality and agency, 47–9, 52–5, 60
  expression of, 47, 108
  grammatical tests of, 44, 55, 120
  intention and, 44–7, 53, 55, 61, 229–30
  logical form of sentences about, 121–2, 125–7
akrasia, *see* weakness of will
Alston, W. P., 277n, 287
anomalous monism, 214–25, 231, 241
Anscombe, G. E. M., xi, 3n, 9, 22, 41n, 59n, 70, 97, 147, 194n
Aquinas, T., 22, 28n, 33, 35, 36n
Árdal, P. S., 281n
Aristotle, 9, 11, 14, 25, 28, 31–3, 35, 39, 41n, 96, 97, 99
Armstrong, D., 78–9
Attfield, R., 241–2
Attwell, J. E., 54n
Augustine, 41

Austin, J. L., 29, 44n, 45n, 56, 57, 65–74, 80–1, 107, 132, 135n
Ayers, M. R., 68–9

Baier, A., 293
Baier, K., 34n
basic actions, *see* actions, primitive
behaviourism, 217
belief:
  as cause of action in combination with desire, *see* primary reasons
  as cause of pride, 279–80, 285–90
  connection with intention, 91–6
  Hume on, 283
  see also decision theory
Bennett, B., xiv, 48n, 78, 136n
Berofsky, B., 160n
Bilgrami, A., xiv
Binkley, R., 291
Black, M., xiii
Bowie, L., 210n
Brand, M., 57n
Brandt, R., 213n
Bratman, M., xiii
Bronaugh, R., 291
Brown, D. G., 291
Brown, S. C., 292
Burks, A., 152
Butler, S., 30, 35

Cargile, J., 137–46
Carlsmith, M., 235
Carnap, R., 41
Castañeda, H.-N., 125–7, 182n, 292
causal explanation, 158–62, 171
  of action, 3, 9–19, 31–2, 47–8, 72–3, 87, 110, 171, 232, 254, 262–75

304 *Index*